Unending Quest

Throughout his existence man has sought the wellspring of life. He has searched for it through superstition, magic, religion, and in modern times, through science. In this sparkling and lucid book, Isaac Asimov routes the path of that exploration, and details its exhilarating discoveries.

He tells how the theory of spontaneous generation was first questioned by the experiments of the Italian Francesco Redi in 1668 with decaying meat and maggots; further debunked by the Dutch Van Leeuwenhoek and his microscope that unveiled minuscular life-forms; and disproved forever by the great Pasteur and his famous experiments.

He traces the theory of evolution from Genesis, through Darwin, to Mendel's findings on genetic inheritance. The history of the cell is also recorded: how one group evolved into the chlorophylls with the ability to turn sunlight into food, and thus freed life from the ocean's scum.

He examines facets of biochemistry, the seeming alchemy by which the elemental substances—carbon, oxygen, hydrogen, nitrogen—convert into the carbohydrates and proteins which are the basis of a continuing life-cycle.

And he returns to the question: when and how did energy strike this cold planet, and like the finger of God, suffuse life into the inert elements of its oceans?

ISAAC ASIMOV

The Wellsprings
* * * *
of Life

E 70

A MENTOR BOOK
NEW AMERICAN LIBRARY
TIMES MIRROR
NEW YORK AND SCARBOROUGH, ONTARIO
THE NEW ENGLISH LIBRARY LIMITED, LONDON

577
A832

To Truman M. Talley, who is Madison Avenue at its best

© 1960 BY ISAAC ASIMOV

Published by arrangement
with Abelard-Schuman, Limited

 MENTOR TRADEMARK REG. U.S. PAT. OFF. AND FOREIGN COUNTRIES
REGISTERED TRADEMARK—MARCA REGISTRADA
HECHO EN CHICAGO, U.S.A.

SIGNET, SIGNET CLASSICS, MENTOR, PLUME AND MERIDIAN BOOKS
are published in the United States
by The New American Library, Inc.
1301 Avenue of the Americas, New York, New York 10019,
in Canada by The New American Library of Canada Limited,
81 Mack Avenue, Scarborough, 704, Ontario,
in the United Kingdom by The New English Library Limited,
Barnard's Inn, Holborn, London, E.C. 1, England

7 8 9 10 11 12 13 14 15

PRINTED IN THE UNITED STATES OF AMERICA

Contents

PART ONE

* * * *

1

The Inevitable Question

The question of the beginning of life almost forces itself on mankind. What child can be so dead to curiosity as not to wonder, on occasion, where he came from, and how? The innocent question is almost traditional. Parents whose children never asked, "Where do babies come from?" would probably feel uneasy, and, I think, rightly so.

Even if a youngster were not dimly aware that he had not always been on the scene and if he were not, therefore, curious (or even apprehensive) concerning his own origin, there would still be the drama of birth all about him. The arrival of a younger brother or sister would be preceded by months of excitement and suspense, the mystery of which he himself would only vaguely share and which would consequently pique and frustrate him. There would be a disquieting and frightening change, both physical and temperamental, in his mother. Finally, there would follow such a revolution in family procedure (usually to his own disadvantage in terms of loss of attention received) that he must brood about it all and, eventually, ask.

And if he remained an only child, there would still be friends who would go through this traumatic experience. A new baby would appear out of nowhere, and the friend would have a possession he himself would not have.

Moreover, this question, "Where do babies come from?" though traditionally asked and rarely unexpected, is also traditionally embarrassing and difficult to answer. Modern mothers may frequently launch into some bowdlerized version of the biological background of birth, but rarely do they do so with poise. And in most cases, even today, the earliest explanations of the process leave the child with the thought that children are found either under cabbage leaves or under a hospital bed, and that they are brought either by a stork or in a doctor's little black bag.

Such explanations would satisfy all but the most formidable youngster, since he would have no reason to suspect that there is anything inherently improbable in the creation of a baby out of nothing. When, later in life (and, perhaps, thanks to the folk wisdom of the gutter, not very much later in life), he learns that the baby originated as a result of the activities of the father and the mother, this activity is what he may find difficult to believe.

But believe it he must, eventually. The life of the baby, he must finally admit, is the product of the life of the parents; human life arises from human life.

If the child is brought up on a farm, he is apt to gain an accurate insight into the process of baby-making much earlier than the city child, since he will undoubtedly have a chance to observe the behavior of livestock on the farm. He will learn soon enough that calves, colts and chicks are the products of their parents, and he will learn in detail the indispensable (if transient) role of the bull, the stallion and the rooster in the process.

Then again, the crop that is laboriously grown and triumphantly harvested springs not from the sterile earth, but from the seeds produced by the crops of yesteryear. Life comes only from life in the case of every animal man herds and of every plant man cultivates.

All this, which each child must discover for himself, either through observation or explanation, with greater or lesser trauma, had to be discovered by mankind as a whole at some early stage of culture. Probably the discovery proved no easier for mankind than for the individual child.

The question, "Where do babies come from?" was often answered by primitive man with tales equivalent to those of the stork and cabbage leaves. In Greek legends, for instance, there are stories of mares which turn their backs to the fructifying east wind and are made pregnant thereby, bearing foals of extraordinary speed. This may have been merely symbolic to later, more sophisticated Greeks, but it may well reflect an early stage where it was actually believed that the wind could be responsible for babies.

The numerous legends of god-born heroes in Greek myths may also reflect the early period when men were honestly uncertain of what brought about pregnancy—perhaps a god, perhaps a ritual prayer, perhaps sitting under a sacred tree. The fertility rites in primitive agricultural societies may have originated, in part, from the same uncertainty.

And the truth, when finally learned, may well have proved as embarrassing for mankind generally as for each child individually. Some people have seen a reflection of this momentous discovery in the biblical legend of the forbidden fruit which Adam and Eve ate and which brought sin and mortality into the world.

Yet, trauma or not, by the time any society had grown sophisticated enough to develop writing (an invention that marks the boundary between the prehistoric and the historic), they had also grown sophisticated enough to know where babies came from. The supernatural and mystic were put aside, and the baby was accepted as the product of the sexual activity of the mother and father. And this, with appropriate modifications, served to explain where lambs, pups, kittens, goslings and fruit tree saplings came from.

It would seem that, having discovered this about man and having made the extension to various plants and animals, it would be simple and easy to make a further extension to all plants and animals, to suppose that all young of whatever kind were the product of parents.

And yet that next step (which to us, out of the wisdom of hindsight, seems so natural) was not taken until modern times.

After all, if we try to put ourselves in the place of our ancestors, we will notice that there are animals and plants which survive despite the fact that they are not cared for by man. They survive, in fact, despite all man can do to wipe them out.

It is exasperating. Useful domestic animals must be carefully guarded and watched over if they are to remain alive and healthy, yet creatures such as mice, rats, mosquitoes and flies flourish and multiply, though unrestricted and merciless war is declared on them. The tender grass is nurtured with love and plant food, while dandelions are poisoned and torn up; but it is the grass that perishes and the dandelions that rise triumphant over adversity.

Where do the vermin and weeds come from?

It is only too easy to fall into the exasperated belief that they spring up from the soil itself; that they are formed of mud and corruption; that their birth, in short, is a kind of conspiracy on the part of inanimate nature to spite man by turning itself into noxious forms of life.

Thus, in *Antony and Cleopatra* (Act II, Scene 7) Shakespeare has the Roman, Lepidus, say, "Your serpent of Egypt is bred now of your mud by the operation of your sun; so is your crocodile."

Lepidus was half-drunk at the time, and even when sober, he was not (as pictured by Shakespeare) a great brain. His drinking companions, Antony and Octavius, who knew better, gravely went along with the gag.

Obviously, Shakespeare himself believed no such thing and introduced the statement as a piece of comedy, but it is quite certain that many in his audience found sufficient humor in the drunken byplay and were quite content, otherwise, to believe that the corrupting mud of the river Nile would indeed bring forth serpents and crocodiles to plague mankind.

The Egyptians themselves (and reasonable foreign observers, such as Herodotus) knew very well that serpents and crocodiles laid eggs and that only from those eggs were new serpents and crocodiles produced.

But then, serpents and crocodiles are sizable creatures and their eggs are large and easily noticeable. Smaller vermin can be more misleading. Field mice may make their nests in holes burrowed into stores of wheat, and those nests may be lined with scraps of scavenged wool. The farmer, coming across such nests, from which the mother mouse has had to flee, and finding only naked, blind and tiny infant mice, may come to the most natural conclusion in the world: he has interrupted a process in which mice were being formed from musty wheat and rotting wool.

Which goes to prove that many a false theory is firmly grounded on the best evidence of all: "I saw it with my own eyes!"

Let meat decay and small wormlike maggots will appear

in it. Eventually those maggots become flies. Out of dead meat come live worms and insects. This is no vague theory. This is eyewitness evidence, as any man can prove for himself with nothing more than a piece of decaying meat.

The greatest and clearest mind of the ancient world, that of Aristotle of Stagira, believed this, as indeed he had to, on the evidence he had. He believed in the ability of non-living matter to give rise to certain types of living creatures as a matter of constant and everyday occurrence. This is called the doctrine of *spontaneous generation.*

This doctrine was accepted and taken for granted by all learned men throughout ancient times, throughout the Middle Ages and into early modern times.

The first crack in the doctrine appeared in 1668, when an Italian physician and poet named Francesco Redi thought he would supplement the evidence of his eyes by arranging an experiment. (By and large, the ancient thinkers were content to observe Nature as it existed and unfolded. They did not try to experiment; that is, to interfere with the natural course of events and thus force Nature to give an answer to some question. This failure to experiment, more than anything else, set narrow bounds to the advance of Greek science.)

Redi noticed that decaying meat not only produced flies but also attracted them. Others before him must have noticed this, too, but Redi was the first to speculate that there might be a connection between the flies before and the flies after; at least he was the first to test such a speculation.

He did this by allowing samples of meat to decay in small vessels. The wide openings of some vessels he left untouched; others he covered with gauze. Flies were attracted to all the samples but could land only on the unprotected ones. Those samples of decaying meat on which flies landed produced maggots. The decaying meat behind the gauze, upon which the foot of fly had never trod, produced no maggots at all, although it decayed just as rapidly and made just as powerful a stench.

Redi's experiments showed plainly that maggots, and flies after them, arose out of eggs laid in decaying meat by an earlier generation of flies. It was just as with serpents and crocodiles, but because the flies' eggs were so small, they went unobserved and so arose the misapprehension.

This was a blow against the doctrine of spontaneous generation, but not a conclusive one. It proved that maggots arose from flies, but after all there were other kinds of vermin, and even Redi himself was willing to believe that some kinds generated with true spontaneity. Besides, just about the time Redi was allowing meat to decay under gauze, a Dutch janitor named Anton van Leeuwenhoek was riding a hobby which raised the whole problem of spontaneous generation to a new level of difficulty.

Van Leeuwenhoek's hobby was the grinding of small lenses, the best and most perfect lenses ever ground up to that time. These could magnify the size of small objects up to two hundred times, and Van Leeuwenhoek went about magnifying everything he could find, from drops of blood to strands of hair, from tooth scrapings to ditch water. In doing so, he made dozens of important discoveries and earned immortal fame for himself.

In 1675 he discovered living things in ditch water that were too small to be seen by the naked eye. The "animalcules" (which we now call *protozoa,* from Greek words meaning "first animals") were only one fifth of an inch or less in length. In 1680 he discovered that yeast, which mankind had been using for ages to make bread with, was made up of tiny living things even smaller than most protozoa. And in 1683 Van Leeuwenhoek observed still tinier living things, which we now call *bacteria* (from a Greek word meaning "little rod").

Where did these microscopic creatures come from? Van Leeuwenhoek learned how to make a broth, by soaking pepper in water, in which protozoa would multiply. Others invented improved broths. But it was not necessary, after all, to go to ditch water for a supply of the creatures. Boil a broth and filter it until there is nothing in it that the lens of a microscope can detect. Wait a while and look again, and the broth is inevitably swarming with life.

Where does it come from? Surely the dead broth has given rise to life. Surely spontaneous generation has occurred. So obvious did this seem that the believers in the doctrine were quite content to see it overthrown in connection with such things as mice and flies. They concentrated on microscopic life.

The first attack on the doctrine at this new level was delivered by an Italian scientist named Lazzaro Spallanzani.

In 1767 he not only boiled broth but then sealed off the neck of the flask containing it. The broth, boiled and sealed, never developed any form of microscopic life. Shortly after the seal was broken, however, life began to swarm.

A sealed neck, keeping out the air, acted like Redi's gauze, and the conclusions had to be similar to Redi's conclusions. There are microscopic and unseen creatures all about us in the air which are smaller and harder to observe than even the eggs of flies. These airborne bits of life fall into any broth left open to the air and multiply amazingly. If they are kept out of the broth, no life originates.

In 1836 a German naturalist, Theodor Schwann, went even further. He showed that broth remained sterile even when open to air, provided the air to which it was exposed had been heated first in order to kill any forms of life in it.

The advocates of spontaneous generation were not silenced by this. Heat, they said, not only destroyed life forms in the broth, but also destroyed some mysterious "vital principle." The absence of this "vital principle" made it impossible for the broth to bring forth life. If air was allowed into the container, it was not forms of life within it but the "vital principle" which brought life into being. Of course, if the air were heated first, *à la* Schwann, the "vital principle" in it was destroyed and it was no help.

About 1860 the French chemist Louis Pasteur knocked that idea on the head once and for all. He maintained that ... scopic forms of life (*microorganisms,* we now call them, including all forms such as protozoa, yeast and bacteria) give rise to disease in man and animals, and that they cause decay and corruption besides. In their absence, he intended to show, organic material would not decay.

He boiled a meat broth until it was sterile, in a flask with a long, thin neck that bent down, then up again, like a horizontal S. Then he neither sealed it off nor stoppered it. He left the broth exposed to cool air.

The cool air could penetrate freely into the vessel and bathe the broth. If it carried a "vital principle" with it, that was welcome. What did not enter, however, was dust and microscopic particles generally. These settled at the bottom of the curve of the flask's neck.

The result? The broth did *not* corrupt; it did *not* breed bacteria; it did *not* show any signs of life. Once Pasteur broke off the swan-neck, however, and allowed dust and particles to reach the broth with the air, corruption set in and life made its appearance.

It was not, then, air that caused life; it was not a "vital principle" contained in the air. It was living creatures that caused life, and if they were kept out of any dead material, that dead material would stay very dead very permanently.

In many cases where a cherished belief is overthrown by scientific inquiry, the Bible is used as evidence by some of the supporters of the old beliefs who are anxious to bring the authority of religion to their aid. Usually, this is done by quoting a verse or verses out of context.

For instance, I could point out that Exodus 8:16-17 states:

> And the Lord said unto Moses, Say unto Aaron, Stretch out thy rod, and smite the dust of the land, that it may become lice throughout all the land of Egypt. And they did so; for Aaron stretched out his hand with his rod, and smote the dust of the earth, and it became lice in man, and in beast; all the dust of the land became lice throughout all the land of Egypt.

By themselves, these verses could be used to support the doctrine of spontaneous generation. Yet, in the larger context, these verses refer clearly to a special intervention of God in human affairs. Because dust is turned to lice as one of the plagues of Egypt, this need not mean that dust turns consistently into lice as part of everyday life. You might as well think that serpents are routinely formed out of dead rods, because earlier in the same book of Exodus Aaron's rod becomes a serpent.

In fact, the Bible seems to speak out clearly against the notion of spontaneous generation. I would like to go into detail about this, not because the Bible is, or ever was intended to be, a scientific textbook, but because the Bible has so often been used to combat some of the ideas of modern science, that I think it would be refreshing to show that it can also be used to support some of these ideas. Besides, the passages involved here have had a particularly powerful influence on the history of biology and will have to be referred to more than once later in the book.

The first chapter of Genesis describes the creation of

heaven and earth, including all the living things on our planet. On the third day of Creation, God is described as creating the plant world. On the fifth day, he created the animal life of the sea and the air. On the sixth day, he created the animal life of the land, including man. (It may surprise some people to learn that no special day was reserved for the creation of man.)

In each case, the individual creation of life specifically involves its further multiplication without further intervention by the deity. In Genesis 1:11, it is written:

> *And God said, Let the earth bring forth grass, the herb yielding seed, and the fruit tree yielding fruit after his kind, whose seed is in itself, upon the earth: and it was so.*

The creation of plant life includes provision for seeds and fruit so that future generations are automatically provided for.

After the creation of sea and air life on the fifth day, God is quoted as saying (in verse 22): "Be fruitful, and multiply, and fill the waters in the seas, and let fowl multiply in the earth."

At the end of the sixth day, God uses almost the same words, addressing them specifically to man, but applying them, by extension, to the remainder of that day's creation, the other animals of the land: "Be fruitful, and multiply, and replenish the earth . . ." reads verse 28 in part.

In fact, even after God grows angry with mankind and decides to destroy them (as recorded in Genesis 6) He makes no provision for a fresh act of creation thereafter. Instead, He makes the initial act in Genesis 1 still hold good by saving one family out of the Flood.

Not only does He save Noah, his wife, their sons and daughters-in-law, but He saves a sampling of every other living creature as well. Genesis 6:19-21, reads:

> *And of every living thing of all flesh, two of every sort shalt thou bring into the ark, to keep them alive with thee; they shall be male and female. Of fowls after their kind, of every creeping thing of the earth after his kind, two of every sort shall come unto thee, to keep them alive. And take thou unto thee of all food that is eaten, and thou shalt gather it to thee; and it shall be for food for thee, and for them.*

Carefully, the living things are enumerated; even the plant world comes under the heading of "all food that is eaten," and presumably seed was preserved for the world after the Flood. Only the animals of the sea are not mentioned and obviously these would survive the Flood anyhow. (Some plant life also survived, perhaps in the form of seeds lying dormant in the temporarily drowned soil, for the dove sent out by Noah came back with an olive leaf in her beak—Genesis 8:11.)

Certainly, on the whole, the biblical account goes to extraordinary lengths to preserve a record of but one single week of Creation and no further creation at any time thereafter.

So strong is this feeling that creation should be left only to God and to only one week of history, that rabbinical legends tell how, in addition to the objects enumerated in Genesis 1, God, in that same original week, also created certain special objects which He planned to use later in world history. Thus, there were the various creatures involved in the plagues of Egypt, including the lice mentioned earlier; there was the gulf in the earth which was one day to swallow the rebellious Korah (Numbers 16: 31-32); the mouth of Balaam's ass, which was one day to talk (Numbers 22:28-30), and so on. This was all to prevent the necessity of imagining God as improvising special acts of creation.

In view of all this, it is surprising that the doctrine of spontaneous generation was never considered irreligious. However, when Pasteur's final destruction of the doctrine came to pass, there was at least no feeling that this new view was irreligious. Other scientific conclusions advanced at the same time brewed storms and tempests of controversy, but not Pasteur's. The notion that (after the original Creation, at any rate) life comes only from life was accepted, after Pasteur's experiments, with virtual unanimity and with complete quiet.

So at least we have made a beginning. The child's question as to where babies come from can be answered "From adults," and such an answer will hold not only for human babies, but for kittens, caterpillars, bean sprouts and bacteria.

2

The Classification
of Creatures

The dictum that life comes only from life is not an entirely unrestricted one. To be more exact, we should say that life comes only from similar life. The new life that arises from the sexual activity of cats is invariably in the form of kittens and never, by any chance, in the form of puppies.

The likeness is never perfect; an individual kitten may be different from other kittens of the same litter in size or coat color; it may be different from either parent as well, in some ways. However, the young of cats possess those general characteristics, despite all minor variations, that spell CAT.

The same can be said of dogs, goats, sparrows, grasshoppers and so on. Each has young like itself; each was born of parents like itself; each comes from a long line (extended indefinitely backward in time) of creatures just like itself. Early naturalists could not but assume, then, that each of these creatures was distinct, as a type, from all others throughout the history of life. And because each had its own distinctive appearance, such separate types of animals were called *species,* a Latin word meaning "outward appearance."

But outward appearance can be considered and yet prove misleading. Think of a Persian cat, with its long, silky fur and compare it to a Siamese cat with its slender body and the startling black markings of its extremities. Persian cats, mated, will yield Persian kittens and not Siamese, while Siamese cats will give birth to Siamese kittens and not Persian. Are these to be considered two different species of animals, as distinct from each other as either is from a dog?

Not so. Neither a Persian nor a Siamese cat can be successfully crossed with any kind of dog. However, a Persian cat will, if given the opportunity, mate with a Siamese cat; and from that mating will arise perfectly healthy kittens possessing some characteristics of each. The relationship between the two cats, despite their differences in appearance, must thus be closer than the relationship of either to a dog. The Persian cat and the Siamese cat are not two different species, but two different varieties within a single species.

The same may be said of the various breeds of dog. A mongrel can be created, by judicious matings, with an ancestry including every known breed.

For that matter, all human beings, no matter how they differ among themselves in outward appearance, can interbreed freely, so that the entire human population of the earth is considered a single species.

On the other hand, consider the elephant. Most people simply say "an elephant" the way they would say "a man"; as though there were one type of animal that can be considered an elephant. Actually, there are elephants that live in Africa and elephants that live in India and southeast Asia and there are differences in appearance among them.

The Indian elephant has a pair of prominent bumps on its skull and a depression at the nape of its neck, both of which the African elephant lacks. The African elephant is somewhat larger and has tremendous ears, while the Indian elephant has comparatively small ears. The African elephant has two finger-like projections at the end of its trunk, while the Indian elephant has one, and so on. All these differences seem minor to most of us, and certainly the two types of elephants are by no means as different in appearance as are a St. Bernard, a greyhound, a Scotch terrier and a King Charles spaniel. It would seem justifiable to consider the two elephants as varieties of a single species. However, this is not so. The two species will not interbreed, and each kind produces only its own and is produced only by its own. They are two separate species.

On the other hand, there are elephants in Ceylon without tusks and elephants in Sumatra with particularly slender trunks. These, however, may be mated with each other and with the elephants in India and Siam. They are only varieties within a species.

In short, the simplest way of defining a species is not on the basis of appearance but on the basis of reproductive behavior. A species is a group of living things which mate successfully only among themselves and which produce young similar to themselves.

(Actually, this is only a rough definition and biologists have despaired of working out an exact one. The horse and donkey, which are usually considered members of different species, can interbreed to produce the mule or the hinny—depending on whether the jackass or the stallion, respectively, is the father. In the same way, a lion and tiger may be crossed, or cattle and the bison. This sort of interbreeding, however, is usually brought about by man's contriving, and would not take place in the natural course of events.)

This notion of difference by reproductive behavior is quite acceptable as long as we deal with familiar creatures. It becomes a little harder to take when we enter the world of creatures with which we are less intimately acquainted (except perhaps as nuisances).

For instance, there is a little verse by Roland Young about the difficulty of distinguishing the he-flea from the she-flea, which ends with the roguish line, "But she can tell; and so can he."

That is a neater trick than the versifier perhaps realizes. She can not only tell a he-flea from another (and, to her, useless) she-flea; she can also tell a he-flea of her own species from a useless (to her) he-flea of some other species. And he can do the same with respect to she-fleas. This becomes something of a feat, when you consider that there are today five hundred recognized distinct species of fleas.

There may seem something mysterious and even magical in the way that a flea can distinguish mating material from 499 varieties of nonmating material, but to a flea there are undoubtedly distinct differences between them. (A flea granted a super-fleaish intelligence might wonder how a man can possibly distinguish a woman from a female chimpanzee; but we can tell, can't we?)

In any case, once the notion of the species arose, the natural question arose sooner or later: "If cats arise only from cats, was there ever a first cat, and if so, how did it arise?"

It was always assumed that there was a first cat. No human culture ever assumed that life had always existed through a past eternity (endlessness being apparently an uncomfortable thought for all but the most sophisticated). And if there were a first cat, it must have been created through some supernatural agency, since there was no previous pair of cats from which it could have arisen in the course of nature.

Throughout the Middle Ages and early modern times, it was taken for granted that Genesis 1 described creation of each species of plant and animal.

In Genesis 1, we find God described as creating "grass, the herb yielding seed, and the fruit tree yielding fruit after his kind . . ." (verse 11); ". . . great whales and every living creature that moveth . . . after their kind, and every winged fowl after his kind . . ." (verse 21); ". . . the beast of the earth after his kind, and cattle after their kind, and every thing that creepeth upon the earth after his kind . . ." (verse 25).

The repetition of the phrase "after their kind" or "after his kind" seems an obvious reference to the species. Each sort of creature formed is to produce young "after his kind," that is, of his own type.

It seemed, then, quite plain that, according to the Bible, all species had been created during that original week. A male and female cat were created, as well as a male and female dog, and a male and female of both the Indian elephant and the Asiatic elephant, and so on. Each species had existed from the beginning and had continued to exist essentially unchanged and separate from all other species ever since. This was the doctrine of the *immutability of species*.

To the average man, through many long centuries, there seemed nothing in the least improbable or uncomfortable about that notion. After all, how many species of creatures are there? Surely not many. (Even today the average man, if asked to sit down and make a list of various species of living creatures, both plant and animal, might find it difficult to think of more than a hundred.)

There is nothing incongruous in imagining a hundred species or so to have been created by God and to have been brought to Adam for naming (Genesis 2:18-20). One could picture Adam naming them all in an hour or so.

Nor was the picture of Noah's Ark particularly troublesome. To bring "two of every sort" of animal into the Ark would not crowd it unduly. The Bible describes it as a ship that was three hundred cubits by fifty by thirty (Genesis 6:15). Since the Hebrew cubit was about seventeen and a half inches, the Ark would be the size of a modern destroyer, perhaps; to the medieval mind this would seem tremendous and ample room for the few animals that had to be carried.

Nor did the learned men of ancient times offer much to disturb this picture. The keenest observer among the Greeks, Aristotle, could list only about five hundred species of animals, and his pupil, Theophrastus, the most eminent botanist of ancient Greece, could list only about five hundred species of plants.

However, as knowledge of hitherto-undiscovered lands increased and observation of nature continued, more and more species of both plants and animals were identified. By 1700 over ten thousand species of plants and animals were known and described, and by 1800 the number had grown to more than seventy thousand.

The larger the number of species, the more grotesque became the literal interpretation of the words of Genesis 1. Even in 1700 the picture of Noah's Ark with all those animals on it had become an uncomfortable one.

As for today—there are now known to be at least one and a quarter million (!) distinct species of plants and animals. Adam is described as naming "every beast of the field and every fowl of the air." Even if this is taken to mean only the mammals and birds, he would have had to name some tens of thousands of creatures.

Noah's Ark would have had to carry millions of creatures, including male and female of each of at least a mil- different insects (500 kinds of fleas alone).

If the words of Genesis 1 are taken literally, they can easily be made to seem ridiculous. And yet what alternative is there? Since cats don't give rise to dogs or dogs to cats, or any species to any other species, there must have been one pair of each species to begin with. Where did they come from? There seemed no way out but Genesis 1.

Yet a way out did slowly appear, and the first steps in the direction of solution arose out of the very flood of species that were being discovered. Naturalists couldn't handle them all separately; they had to find ways of grouping them, if only to avoid being drowned by them.

The idea of classification is not a difficult one to get. Each species is not equally different from all other species. There are greater and lesser similarities which even a child can see. A child of three, seeing a tiger at the zoo for the first time, is very likely to point his finger at the creature and say "Pussycat."

A tiger is not really a pussycat, of course, and to mistake one for the other could easily lead to disaster. Nor is there any doubt as to the fact that the two creatures are of distinct species. There is no more chance of crossing a pussycat and a tiger than of crossing a pussycat with a pussy willow.

And yet the child is, in a way, perfectly right. The tiger looks, unmistakably, like a giant cat. Lions, panthers, lynxes, jaguars, ocelots, ounces and cheetahs all look like large cats and it takes practically no effort at all to refer to them (each a fully separate species of living thing) as belonging to the "cat family."

You can easily spot the wolf, fox, jackal and coyote as belonging to the "dog family." You can, without trouble, speak of a "bear family," a "rat and mouse family," and so on.

Or you might make a larger classification by speaking of "four-footed beasts" or "quadrupeds," which would include all the cats, dogs, bears and rats; or the "feathered creatures," or the "creeping things."

It is important to remember that while the species is a more or less natural unit based on the reproductive behavior of the organisms themselves, any system of classification by which different species are grouped into larger units is strictly man-made. The nature of the classification depends upon what strikes the individual classifier as logical; as circumstances vary, so do the conclusions of logic.

You might want to speak of the cat family and lump the pussycat with the lion. A naturalist would have good reason to do so. You might also speak of household pets and lump the pussycat with the canary. And where it is a matter of feeding an animal by hand, it is much better to lump the cat and the canary than to lump the cat and the lion.

A very traditional method of classification is to divide species into groups that share a particular habitat. We do it ourselves when we talk of "land creatures" and "sea creatures." The Bible does essentially the same.

For instance, in Genesis 1:21, when God created the denizens of the sea, it is particularly stated that "God created great whales." Whales were thus lumped with the finny creatures of the deep, as well as with oysters and lobsters. The general term for all these creatures was "fish" and today we still speak of "shellfish" and of "starfish," although these are far more different from cod and mackerel than we ourselves are. If a fish is to be defined as "any creature that lives in the sea and not on the land," then a whale is a fish.

The Greek philosopher Aristotle, however, noticed that porpoises and dolphins (small members of the whale family) breathed through lungs, not through gills as ordinary fish did. What is more, porpoises and dolphins did not lay eggs as ordinary fish did, but brought forth live young. And as a result of remarkably careful observation, he noticed that the young porpoises were nourished before birth by means of an organ within the mother's body which we today call a placenta. Thus, the whale family is completely unlike ordinary fish and exactly like the hairy quadrupeds of land.

To Aristotle, the manner in which the whale family brought forth young was more important than the mere fact that whales lived in water and had a fishlike shape. He therefore classified whales with the quadrupeds and not with the fish.

(He was a voice crying in the wilderness in this respect. For two thousand years after Aristotle the whale remained a fish in the minds of men, and only in quiet modern times has Aristotle's view won out. Nevertheless, it is not wrong to say a whale is a fish. It depends on the system of classification. If nowadays we call the whale a mammal and lump him with mice, porcupines and giraffes, it is only because to do so is more useful and better fits the purposes for which modern scientists use their systems of species classification.)

In the same way, in Leviticus 11:19, the Bible specifically lists the bat among the birds, because it is a flying creature that inhabits the air. We ourselves notice that ordinary birds all have feathers and lay eggs, while the bat has hair and brings forth live young with the preliminary help of a placenta. We find it more useful, therefore, to restrict the definition of birds to feathered inhabitants of the air (and to feathered creatures that cannot fly, also, for that matter) and to consider the bat a mammal.

In early modern times, a series of naturalists attempted to make some sort of systematic classification of all species known to them. An outstanding example was an Englishman named John Ray (or Wray) who, beginning in 1660, began to classify more and more plants and eventually listed about 18,600 species of plants, neatly divided into groups according to a system that to him seemed logical.

For instance, he divided flowering plants into two groups, according to whether the seeds contained one tiny little leaf or two. These little leaves lay in the seed in a small hollow resembling a type of cup the Greeks called a *kotyle*. The leaves were therefore called *cotyledons,* and Ray referred to his two groups as *monocotyledonous plants* and *dicotyledonous plants*.

This particular type of classification proved so useful that it is still used today. You may wonder that the difference between one embryonic leaf and two should prove so important. In itself, of course, the difference is not important. Observation of the plants showed, however, that there were a number of other ways in which all monocotyledonous plants differed from all dicotyledonous ones. The difference in leaves was just a handy marking, so to speak, which was symptomatic of many general differences.

(Naturally, you have to be careful in picking your visible signs. Not all, no matter how distinctively they seem, will serve as a useful criterion for classification. To define birds as "two-legged creatures" would make a man into a bird. To define them as "winged creatures" would make bats and flies into birds. To define them as "feathered creatures," however, is fine. Experience has shown that it is convenient and involves no uncomfortable contradictions to consider all creatures with feathers to be birds, and no creature without feathers to be a bird. In this connection, I might add that an eighteenth-century naturalist once suggested that man could be defined, in briefest form, as a "featherless biped." The French satirist Voltaire at once pointed out that this would make a man out of a plucked chicken—an example of the dangers of classification.)

In 1693 Ray attempted to classify animals as well; in this he was strongly influenced by a system of classification used in the Bible.

In Leviticus 11 rules are given for dividing the animal kingdom into "clean" (fit for food and for ritual sacrifice) and "unclean" (unfit for these purposes). Verse 3 reads: "Whatsoever parteth the hoof, and is clovenfooted, and cheweth the cud, among the beasts, that shall ye eat."

There is here a threefold classification. First, beasts are divided into those with hoofs and those without hoofs. The latter (such as tigers or anteaters) would be unclean. Then, of those animals with hoofs, there is a division into those with cloven hoofs, that is, with two hoofs per foot; and those with fewer or more hoofs. Thus, the horse (with one hoof per foot) and the rhinoceros (with three hoofs per foot) would be unclean.

Finally, the cloven-hoofed creatures are divided into ruminants (those that regurgitate food that has been swallowed in haste and reswallow it at leisure, a phenomenon referred to as "chewing the cud") and nonruminants. The latter, including swine, are unclean. The former, including cattle, sheep, goats and deer, are clean.

Here, you see, we have a classification based on anatomy and physiology. Anatomy is the study of the structure of an organism and its parts, and the question of the presence and form of hoofs comes under that heading. Physiology is the study of the functions of an organism of its parts; the question of whether or not a digestive system is adjusted to rumination comes under that heading.

John Ray followed the Bible in classifying mammals, but went a bit further. Thus, he divided mammals into two main groups, those with hoofs and those with claws. He subdivided each according to how many hoofs or claws were present on each foot, so that, for instance, there was a subdivision that included the five-clawed mammals. These were further divided into those with narrow claws and those with broad claws (the latter including monkeys, apes and, of course, man).

The double-hoofed animals, Ray divided into ruminants and nonruminants as the Bible did. He went one step further, however, and divided the ruminants into those with permanent horns (cattle, sheep and goats) and those with horns that were shed every year (deer).

Some of the Ray classification still persists today, but a generation after Ray there arose another naturalist who did the job of classification so thoroughly and well that not only have the main features of his scheme been retained ever since, but the work of all his predecessors has faded to insignificance in comparison. The science of *taxonomy* (that is, the classification of plants and animals, from a Greek word meaning "arrangement") is, in its modern form, the creation of Carl von Linné, a Swedish botanist better known to posterity by his Latinized name of Carolus Linnaeus.

Linnaeus began, like Ray, with the classification of plants. He traveled throughout Scandinavia (including unexplored regions in the far north) and other parts of Europe in order to observe many species carefully.

In 1737, at the age of thirty, he wrote his greatest book, *Systema Naturae,* in which he attempted to classify all the known species. In principle, he did only what Ray and other predecessors had tried to do, but Linnaeus went further in choosing just those characteristics which served best to differentiate the various groups. He described each species succinctly and well. And he went further than anyone else in systematically building up groups of species, then groups of groups, then groups of groups of groups.

Thus, similar species were grouped into a *genus* (plural, *genera,* from a Latin word meaning "race" or "sort"). Similar genera were grouped into an *order* and similar orders were grouped into a *class.* Linnaeus divided all animals into six classes: mammals, birds, reptiles, fish, insects, and worms. After Linnaeus, this system was carried even further. Similar classes were grouped into a *phylum* (plural, *phyla* from a Greek word meaning "tribe"), and similar phyla were grouped into a *kingdom.* There are now two kingdoms of life generally recognized, those of the plants and the animals.

The exact divisions of a kingdom into phyla, or a phylum into classes, and so on, are never universally agreed upon. Taxonomists will usually disagree whether a particular group of animals is so different from others that it belongs in a phylum by itself, or whether it should only form a separate class within an already well-known phylum. The same is true all the way down, so that there will be arguments as to whether a number of species ought to be included in one genus or divided among two or more genera. None of these disagreements affects the main outlines of the present classification of living species; they are only quarrels over relatively minor details. I mention them only to point up the fact that classifications are not hard and fast even today; in fact, according to modern thinking, there is good reason to suppose that classifications cannot ever be hard and fast. We will come back to this later.

Another of Linnaeus' contributions was to popularize the custom of referring to each species by a double name: that of the genus to which it belonged, followed by the name of the species itself. This system has been followed ever since, as has Linnaeus' habit of using only Latin names for the purpose (Latin having long been the language of scholarship in western Europe).

For instance, the Latin words for "cat," "dog," and "elephant" are, respectively, *"felis," "canis,"* and *"elephas."* The *binomial nomenclature* used for some members of the cat family include *Felis domesticus* (the ordinary pussycat), *Felis leo* (the lion), *Felis tigris* (the tiger) and *Felis pardus* (the leopard).

For the dog family, there is *Canis familiaris* (the dog), *Canis lupus* (the European gray wolf) and *Canis occidentalis* (the American timber wolf).

The two elephant species are *Elephas maximus* (the Indian elephant) and *Elephas africanus* (the African elephant). Some taxonomists prefer to put the two elephants in separate genera, so the African elephant may be called *Loxodonta africanus,* the word *"Loxodonta"* being the Latin for "oblique teeth."

This principle of binomial nomenclature is exactly that used by the telephone directory in distinguishing Anderson, Walter, from Anderson, William.

In many cases, the original name suggested by Linnaeus is still in use. For instance, the species to which man belongs was named by him *Homo sapiens* (which in Latin means "man, wise"). It is still used, for all that it may just possibly be a misnomer.

The system that has developed out of Linnaeus' scheme

has the advantage of reducing seeming chaos into remarkable order. It has the convenience of putting every species into its special slot and showing its similarity to other species. It places each species near those species greatly resembling it in anatomy and physiology; further from those resembling it less; still further from those resembling it still less, and so on.

This alone would be enough to please any scientist with a sense of order, and Linnaeus had his own sense of order developed to an almost pathological degree. For him, classification was an end in itself, and he sought for no higher meaning in what he had created.

Not so for others. An orderly presentation was fine, but once the order was established, was there nothing more? Could there not be new truths, hitherto obscured by disorder, now emerging in full clarity?

For instance, almost anyone studying the Linnaean system must be struck by its resemblance to a "tree of life." Imagine the trunk of his tree representing life itself and branching into two major limbs, representing the two kingdoms, the animal and the plant. Each major limb divides further into smaller branches representing the phyla. Each of these splits up into still smaller branches, and so on, until hundreds of thousands of final twigs represent the individual species.

As in a real tree, some of the larger branches divide and redivide prolifically, these representing large groups with many flourishing genera. Other branches seem to be withered, subdividing infrequently and ending with but a few obscure species.

One can even arrange the limbs, branches and branchlets in such a way that in the center of all, at the very tiptop of the tree, is an upthrusting twig which can be labeled proudly *Homo sapiens*.

Anyone drawing such a tree (and many have done so) probably cannot help imagining it as growing and developing, as actively branching out. Many looking at a figure of such a tree must have wondered if in actual fact it was necessary to consider every species, every little twiglet, to have been in existence from the beginning. Suppose various groups of species began in the form of a common ancestor which, with time, branched out into the various genera and individual species, a branch developing branchlets and twigs.

Or suppose that the tree itself had originated as a shoot, so to speak, a little bit of original, undifferentiated life. And suppose that, with time, the whole tree had developed from the largest limb to the smallest twigs.

If people did think this (and some simply must have), they turned thoughts into safer channels and forbore to make a fuss about it. And yet there must have been enough expression of this thought for Linnaeus himself to deny emphatically that his arrangement was to be taken for any indication that one species could develop from another. He was quite categorical about it. There were as many species in existence in his day, he declared firmly, as were created originally by God, not one more nor one less.

Nevertheless, Linnaeus' system of classification proved more powerful than Linnaeus. The thought that related species had developed out of common ancestral forms and were not necessarily separately created, persisted and would not be downed. And, eventually, it won out.

3

The Tree
Comes to Life

What kept the doctrine of the immutability of species alive so long was that there was a major flaw in any theory that one species could develop from or change into another. The flaw was that no one had ever seen it happen. Cats remained cats and catfish remained catfish. If there were changes, they were so small that in the entire course of recorded history no development of a brand-new species had been recorded.

It seemed quite certain that if species changed, it was only at a very slow rate. The development of a new species, if it were possible at all, would require at least tens of thousands, perhaps hundreds of thousands, of years.

And there, indeed, was the rub. In the time of Linnaeus, and for fifty years afterward, men's ideas on the length of time available were still fixed by the literal words of the Bible.

Theologians had devoted considerable effort to tracing back the dates of various biblical events through statements in the Bible as to the length of the reigns of kings, or its statements of the number of years from the time of the Exodus under Moses to the time of the building of the Temple by Solomon, or its statements as to the age of the patriarchs. All these put together brought them to the date of Creation itself, thus setting an upper time limit for all the years available for species to change; and that upper time limit was far too small.

The most familiar conclusion (and one that is still printed in many editions of the King James Version of the Bible) was worked out by Archbishop James Ussher, an Irish theologian of the Church of England, about 1630.

stroyed in a universal catastrophe, and that there had been a number of these, of which Noah's Flood was the most recent, and that there might be more in the future. After each catastrophe there were always remnants of the old forms that survived, and these remnants were changed so that they no longer resembled the fossil forms, but were developed more highly.

However, the study of fossils did not become a true science (now called *paleontology*, from Greek words meaning "the study of ancient living things") until 1791. In that year an English land surveyor named William Smith was working on excavations in connection with canal-building, and noted that rocks were formed in definite layers. These rock layers are called *strata* (singular, *stratum*, from a Latin word meaning "spread out"). Each stratum had its own characteristic fossils, which did not appear in other strata. Furthermore, a particular stratum which sank out of view at one place might crop up again miles away, still with only its characteristic fossils and no others. Eventually, in fact, strata could be identified by the fossils they contained.

What Smith showed was that fossils were not random phenomena but displayed an orderliness which gave good hope for further understanding. (The establishment of order out of apparent disorder is usually the first step leading to a new and deeper insight.)

This sense of order was intensified by the second great name in paleontology, the French naturalist Georges Cuvier. His specialty lay in comparing the anatomy of one species of animal with another, observing both similarities and differences. He is thus considered the founder of the science of *comparative anatomy*. In doing so, he sharpened and extended the Linnaean system of classification. In fact, it was he who first introduced the notion of phyla.

With a background such as this, Cuvier turned his attention to fossils in about 1800, and saw at once that they fitted into his system of classification. Specifically, they fell into one or another of the phyla which he had himself described. They made up part of the scheme of life. Furthermore, the older the fossils (that is, the older the strata in which they were found), the more they differed from the living representatives of their phyla, and, in general, the simpler and less highly developed they were.

From our viewpoint today, with the advantage of hindsight, we may be amazed that Cuvier did not deduce that forms of life had slowly developed from the simpler to the more complex, and that the fossils marked various stages in the process. Cuvier, however, could not free himself of the words of Genesis. He could not deny the great age of the earth, for the formation of the strata in which the fossils were buried, layer on layer, obviously took long eons. And yet, the literal words of Genesis had to be preserved.

What he did was to take up and modify Bonnet's notions of catastrophes. Cuvier suggested the occurrence of at least four catastrophes, each of which wiped out all life and each of which necessitated a new beginning with a completely new creative act. If this view is taken, then Genesis can be considered to refer only to the most recent creation, the one in which man was formed. The previous creations, since they did not concern man, were not mentioned in the Bible. (It was as though the earth were a background used by an economical God who experimented with creation now and then but was reluctant to construct a brand-new world each time.)

Cuvier was by far the most famous biologist of his time, and when he supported "catastrophism," the uniformitarian doctrine of the virtually unknc   Hutton seemed dead. But then another British geologis. Charles Lyell, took up the fight. He was a skilled writer and popularizer, rather than a great originator, and he proved to be the man for the job. In 1830 he published the first volume of a three-volume work entitled *Principles of Geology*. In this he summarized with great clarity all the geological evidence he could find in favor of Hutton's uniformitarian theory. His summary was so logical and convincing that it won out almost at once, and the theory of catastrophes underwent a catastrophe of its own and was killed.

(The simplest argument against catastrophism is this: as knowledge of fossils increased, it became obvious that many ancient creatures endured for long ages without much change. Some creatures actually living today, such as the king crab and a shelled sea creature called Lingula, have been in existence, virtually unchanged, for many millions of years according to the clear fossil evidence. Such organisms are, in fact, often called "living fossils." There is no place in the past history of the earth where a line of catastrophe can be drawn and where one can say that all fossils after it differed from all fossils before. Every possible line would be straddled by the continuing existence, before and after, of many varieties of fossils. So one must conclude that the story of life is continuous and uninterrupted, and though many species and genera—and even whole classes—came to an end, life itself, having once begun, has never once yet come to an end.)

By the middle of the century, then, it was recognized that life on the earth had undergone a long history of at least many millions of years, and that the fossils were the record of that history.

The time was ripe to bring the Linnaean tree to life. Or, if you like, it was time for the development of some systematic theory, that would describe an unrolling, so to speak, of a scroll of life. Life, instead of being discontinuous, like the pages of a book with each page an eternally separate species, was one long, continuous sheet of a large scroll, at the beginning of which were very simple, microscopic creatures. These grew more complicated as the scroll unrolls until, at the point now reached, there are to be found man and the other species now in existence. Such a view would represent a theory of *evolution* (from a Latin word meaning "unrolling").

Merely to speculate about evolution is easy. Even some of the Greek thinkers did that. In order to have such speculations carry weight, however, some logical reason must be advanced to explain why evolution should take place. It is one thing to say that a primitive cat creature might develop into lions, tigers, leopards and so on, and quite another to suggest why it should do so, when it would be so much easier for it simply to remain a primitive cat creature, just as king crabs remained king crabs over many millions of years.

The first to advance reasons for evolutionary change was the French naturalist Jean Baptiste de Lamarck. He began by studying the invertebrate animals and improving their system of classification over that of Linnaeus and even of Cuvier. This set him to thinking how the creatures seemed to change step by step as one went up the classification. In 1809 he published a book entitled *Zoological Philosophy,* in which he described his theories of how these changes must have come about.

Organisms, suggested Lamarck, made much use of certain portions of their body in the course of their life, and under-used others. Those portions that were used developed accordingly, while the others withered. This development and withering were passed on to their descendants.

Lamarck used the giraffe for his most often-quoted example of this. (The giraffe had just been discovered and its queer, distorted shape had been receiving much publicity, naturally.) For instance, he said, a primitive, giraffe-like creature which was fond of browsing on the leaves of trees would stretch its neck upward with all its might to get all the leaves it could. It would stretch out its tongue as far as it could. It would stretch its legs, too. In doing so, its neck, tongue and legs would become slightly longer than they would have been without this practice. The longer neck, tongue and legs would be passed on to the young produced by the creature. When these had grown to adulthood, they would have a longer neck, tongue and legs to begin with, would stretch them more, pass on a still longer one to its young and so on. Little by little, the modern giraffe, with its odd bodily form, would develop from a creature that originally might have had the ordinary proportions of an antelope.

This is an example of the *inheritance of acquired characteristics,* and although it is an attractive theory and sounds very logical, it foundered on the rock of several facts. In the first place, Lamarck's theories postulated an inner drive, a kind of attempt on the part of the organism to evolve. This can be imagined in the case of long necks, with the animal trying to reach the barely reachable. But how would an organism, according to this system, develop protective coloration?

There are many insects that have the color and shape of leaves or twigs; there are fish with a color pattern that makes them fade into a pebbly background; there are striped and splotched animals, like tigers, leopards, zebras and giraffes, that are quite unnoticeable against the splotchy background of sunlight shining through leaves. How did these characteristics develop? Surely a giraffe did not try to be splotchier and pass improved splotches to its youngsters.

Besides, and even more conclusively, acquired characteristics are simply not passed on to the young. A long line of athletes will encourage their young to be athletic in line with the family tradition, but if the young choose to adopt a sedentary vocation, they will be as flabby as though all their ancestors had been sedentary. Some experimenters tried to test whether acquired characteristics were inherited. One cut off the tails of generation after generation of mice, and proudly announced that each new generation was born with complete tails that had to be cut off in turn. None inherited taillessness, nor did the tails grow even slightly shorter with each generation.

This has always been mentioned as an example of a completely foolish experiment, for had the experimenter thought a little, he would have remembered that Jews (and others) have systematically cut off the foreskins of young male children for perhaps a hundred and twenty generations in the rite of circumcision, and yet Jewish children are not born circumcised. They are not even born with undersized foreskins.

But now the English naturalist Charles Darwin steps upon the scene.

In 1831 the young Charles Darwin (he was then only twenty-two) joined the crew of the "Beagl-." This was a ship making a five-year voyage about the world to explore various coast lines and to increase man's knowledge of geography. Darwin went along as ship's naturalist, to study the forms of life in far-off places.

This Darwin did extensively and well, and upon the return of the "Beagle" he wrote a book entitled *Zoology of the Voyage of the "Beagle,"* which was published in 1840 and first made him famous in the world of science.

Much more important, his observations on that most famous voyage in the history of biology set him to thinking deeply about evolution. He had read Lyell's book on geology shortly after it had been published. A friend had shown it to Darwin and warned him that it was amusing but completely harebrained. Darwin found it not harebrained at all. He was converted to uniformitarianism and to the belief in the antiquity and continuity of life. That set the stage for his evolutionary thinking. (After his return from the "Beagle" expedition, he grew personally friendly with Lyell and for a while served as secretary of the Geological Society.)

During the voyage, Darwin noticed how species changed, little by little, as he traveled down the coast of South America, for instance. More striking still were his observations on the Galápagos Islands. These are a group of a dozen or so islands about 650 miles from the coast of Ecuador. The most unusual life forms present on them are the giant tortoises (indeed, the name of the island group comes from the Spanish word for "tortoise").

What Darwin mainly noticed on the islands, however, was a group of birds (called, to this day, "Darwin's finches"). These were closely similar in many ways, but were divided into at least fourteen species. Not one of those species existed on the nearby mainland or anywhere else in the world.

Now why should this be? Why should fourteen species exist on these islands and nowhere else? The islands, it seemed to him from their structure (following the Hutton-Lyell theories), were volcanic outcroppings. They had been built up out of the ocean floor and had never been connected with South America. To begin with, there must have been no life at all upon them until they were gradually colonized by life forms that could reach it from the mainland, by flying, swimming or floating.

In years long past it might have been that a few seed-eating ground finches, of a type found on the neighboring mainland, flew to the islands and multiplied there. Gradually, they had varied as they multiplied. None of the original mainland species was left, only varied species descended from the original. Three of the descendant species were still seed-eating when Darwin found them, one being rather large in size, one medium, and one small, each feeding on its own kind of seeds. Two other species had learned to feed on cacti, and most of the others fed on insects.

Members of one species could not interbreed with members of another, nor did they interfere with each other's livelihood, since each species had a different manner of feeding. (In fact, the easiest way to tell one species from another was by the shape and size of the beak, which in every case suited the particular mode of feeding of that species.)

But, supposing all this history of the ancestral finch and its descendants were correct and just as Darwin imagined it, there remained the question: why should it have happened? Why should not the ancestral finch have remained the same seed-eating species it had been for so long? After all, it had remained a seed-eating species of the ancestral form on the mainland. It had not changed there.

The beginnings of an answer came to him in 1838 when he first read a famous book entitled *An Essay on the Principle of Population,* written by an English clergyman named Thomas Robert Malthus and published in 1798. Malthus maintained that human population always increased faster than the food supply did and that eventually population had to be cut down by either starvation, disease or war.

Darwin thought at once that this must hold for all forms of life and that those of the excess population who were first cut down were those who were at a disadvantage in the competition for food.

For instance, those first finches landed on the Galápagos Islands must have multiplied unchecked to begin with and outstripped the supply of the seeds they lived on. Some would have to starve, the weaker ones first or those less adept at finding seeds. But what if some could turn to eating bigger seeds or tougher seeds or, better still, turn from the eating of seeds to the eating of insects? Those who could not make the change would be held in check by starvation, while those who could, found a new, untapped food supply and could multiply rapidly until, in turn, their food supply began to dwindle. Then some might specialize in larger insects or develop a woodpecker bill to get grubs from under bark that the others could not touch; these would multiply again.

In other words, creatures would adapt themselves to different ways of life, radiating outward from the original form to fill new niches in the environment. This is called *adaptive radiation.*

On the South American mainland, the original seed-eating ground finch could not do this, since the niches were already filled by other creatures. A ground finch with a woodpecker bill would be competing with real woodpeckers that had been much longer at the game, so to speak, and the finch woodpecker would not survive. On the Galápagos Islands, which were virtually empty of life, the niches were there for the taking, and the finches took them as the pressure of overpopulation forced them out of the old niches.

How did these changes come about? How could a seed-eating finch suddenly learn to eat larger seeds which others could not, or learn to eat insects?

Here Darwin was on rough ground, and his answer to that question remained the weakest point of his theory. He decided that every new generation varied randomly from the average. In any litter of pigs, some are larger than others. In any litter of kittens there are trifling variations in color pattern. Similarly, there are variations in everything else. It was by taking advantage of such natural, random variations, Darwin thought, that man has through many generations developed larger and stronger horses to carry man in armor, or faster horses to win races, or cattle to give more milk or more beef, sheep to give more wool, hens more eggs, turkeys more white meat, or cats and dogs to show odd and amusing shapes.

Could not natural forces also take advantage of this innate capacity for random variation?

Suppose, for instance, that two groups of a single species are separated geographically. One group of dogs is taken to Alaska to live, while another group is taken to Mexico. The Darwinian view is that in both groups of dogs the different puppies of a litter would vary in the nature of their fur. Some puppies would be born with the ability to develop fur that was thicker and longer than average; others with the ability to develop fur that was shorter and sparser than average.

In Alaska, those puppies with thicker, longer fur are better protected against the cold. On the average (though not necessarily in particular individual cases) the thick-furred specimens would live longer than the others, since they would be less susceptible to the cold. They would hunt and find food more efficiently, have time and energy to mate oftener and would give birth, therefore, to more, and possibly healthier, young. The sparsely-furred dogs would be at a disadvantage at every point. They would survive with greater difficulty, die sooner, and mate less often. Future generations would descend more from the thick-furred dogs than the sparse-furred dogs; this fact is referred to as *nonrandom mating*.

The litters of thick-furred dogs would be thicker-furred in general than those of thin-furred dogs and among these litters some would be thicker-furred, again, than others, and survive best. From generation to generation, there would be nonrandom mating in favor of thick fur until a shaggy dog like the Alaskan Husky developed.

In Mexico, on the other hand, it is the thin-haired specimen that would get along better. Without the unnecessary weight of fur and the continual discomfort induced by unnecessary heat conservation, it would be stronger, survive more efficiently and mate more frequently. A hairless variety, such as the Mexican Chihuahua, would thus develop.

If the two groups of dogs were separated long enough, change after change might take place as a result of selection among the different varieties by the rigors of the natural environment (this is called *natural selection*). There would not only be differences in temperature, but differences in the food supply available, and so on. No one change would be big enough to separate the two groups into separate species, but eventually the cumulative effect of the changes would do so.

By Darwinian notions, the giraffe got its long neck not because it tried for one (as Lamarck had it), but because some giraffes were born with naturally longer necks and these got more leaves and lived better. In the long run they left more descendants. It was the forces of nature that lengthened the neck, and not an inner drive. This view explained the giraffe's blotched coat just as well. The better the blotches, the less noticeable the giraffe was against a blotched background and the more likely it was to escape the eyes of a prowling lion. Consequently, the more descendants it left.

Some creatures in which the male had very conspicuous coloring (as in the case of many birds) resulted from the female of the species regularly choosing for generations the most flamboyant male as a mate, allowing spectacular appearance to sway her heart more than modest worth, perhaps (a point of view with which *Homo sapiens* is not unacquainted). The development of the peacock, for instance, is the result of *sexual selection*.

It is because the formation of species, whether by natural selection or by its variant, sexual selection, is a continuous and very slow process, that taxonomists have difficulty in deciding whether groups of organisms are one species or two, or whether a group of species fall into one genus or two, and so on. If every species had been independently created, there would be some hope that enough differences would exist between any two of them to enable a clear-cut definition of "species" to be made, one which would draw the line sharply between any two neighboring species.

In the Darwinian view this is impossible. Slowly, and by the gentlest stages, two varieties within a species would drift apart and become separate species, but at no specific point could one point the triumphant finger and say, "Here, right here, two species have formed." There always remains the misty borderline which it may take a developing species tens of thousands of years to cross.

Considering the number of species in existence, there must be hundreds of species just at this borderline, and not all the ingenuity of man can decide definitely whether these particular groups of animals are one species or two.

And as species continue to separate under the rigors of natural selection, neighboring species may differentiate to the point where they will occupy two genera (according to the criteria set up by taxonomists). There again, there will be intermediate stretches of time where it will be impossible to say whether they have reached the two-genera stage or not. No matter what criteria are set up by taxonomists to govern the decisions as to species, genera, orders, classes and phyla, there will always be indefinite intermediate stages.

In short, the very imperfection of taxonomy, the very uncertainties of classification, are strongly in favor of the Darwinian view rather than with the views held by the literal interpreters of Genesis.

Slowly Darwin collected data to show the workings of random variation and natural selection. For instance, there was the question of *vestiges*. Various animals possessed remnants of tissue that were useless (or even harmful) but were the remains, perhaps, of organs that had been useful in some ancestral form. They were the footprints, so to speak, of what had gone before (the word "vestige" comes from a Latin word meaning "footprint").

For instance, whales and snakes have useless scraps of bones that might once have formed parts of hip girdles and legs, showing that they were descendants of creatures that had walked on all fours. A horse has a single line of bones down its leg ending in the one hoof it possesses on each leg. But to either side of that line of bones are two thin splints that come to a dead end, but which once ended in hoofs, showing a horse to be descended from a three-hoofed creature. The kiwi is a flightless New Zealand bird that seems to have no wings (it is also called the apteryx, which in Greek means "no wings"), but close investigation shows small structures hidden under the feathers that look as though they had once been wings but had shriveled away. The kiwi had long-dead ancestors that could fly.

Finally, Darwin started a book on the subject in 1844, but so ardently did he continue to multiply his examples and tighten his reasoning that in 1858 he was still working on it. His friends knew what he was working on and several had read his preliminary drafts. They urged him to hurry or someone else would get there first. Darwin, however, was not to be hurried—and someone else did get there first.

The man was Alfred Russel Wallace, another Englishman, fourteen years younger than Darwin. Wallace's life was quite like that of Darwin: he, too, early became interested in nature and joined an expedition to distant lands.

He traveled to tropical South America and also to the East Indies. In the latter, he noticed that the plants and animals living in the eastern islands of that group (continuing on down into Australia) were completely different from those in the western islands (continuing on up into Asia). The line between the two types of life forms was sharp, curving between various islands, running between the large islands of Borneo and Celebes, for instance, and between the small islands of Bali and Lombok further to the south. That line is still called "Wallace's Line" to this day.

Now the mammals in the eastern islands and in Australia were distinctly more primitive than those in the rest of the world. Could it be that Australia and the eastern islands had early split off from Asia at a time when only primitive mammals of the Australian type existed? There in Australia they flourished, but in Asia new and more advanced mammals developed with which the primitive ones (in Asia) could not complete. In Asia, the primitive forms died out. Wallace's Line would, therefore, be the dividing point that marked the limit beyond which the newly developed, advanced mammals of Asia could not cross. It was the moat that saved the primitive forms in Australia.

But how did the advanced life forms develop? Wallace first began puzzling over this in 1855, while in Borneo. In 1858 Wallace, too, came across Malthus' book, and from it he drew the same conclusions as Darwin.

But there was this difference between Darwin and Wallace. After fourteen years Darwin was still working on his book. Wallace was not that type. Once the idea was clear in his mind, he sat down to write and was finished in two days.

And to whom did Wallace send the manuscript for consideration and criticism? Why, to the famous naturalist Charles Darwin, of course.

When Darwin received the manuscript he was thunderstruck. It expressed his own thought in almost his own language.

However, Darwin proceeded to behave like the ideal scientist. Although he had been working so long on the theory (and, of course, had witnesses to prove it), he did not try to suppress Wallace's work and keep all the credit for himself. He passed on Wallace's work at once to other important scientists, and offered to collaborate with Wallace on papers summarizing their mutual conclusions. This was done. Work by both men appeared in the *Journal of the Linnaean Society* in 1858. (Darwin deserves the lion's share of the credit, just the same. Wallace may have had the theory, but that was the easy part. It was Darwin who gathered the infinite detail of evidence to support the theory.)

The next year Darwin finally finished his book. Its full title is *On the Origin of Species by Means of Natural Selection, or the Preservation of Favoured Races in the Struggle for Life*. We know it simply as *The Origin of Species*.

The learned world was waiting for the book. Only 1,250 copies were printed and every copy was snapped up on the first day of publication. More copies were printed, and they were quickly bought, too. In fact, *The Origin of Species* is an example of a scientific book so well written and organized that it is worth reading as a "classic" for its own sake. Even now, a hundred years later, although Darwin's theory has in many respects been modified and improved upon, the book is still in print. In fact, a Mentor paperback edition has been published in honor of the centennial of the first edition.

As you can easily imagine, Darwin's book and his theory of evolution by natural selection broke on the world (and not just the scientific world) like a thunderbolt. It set up a controversy that has not entirely died even now, a hundred years later.

4

The Evolution
of Evolution

Much of the furore over Darwin's theory arose over its application to man. Lyell, whose geological views had so influenced Darwin, now returned the compliment. In a book entitled *The Antiquity of Man,* published in 1863, he came out strongly in favor of Darwin's theory and discussed the hundreds of thousands of years during which man (or manlike creatures) must have existed on the earth. He used as his evidence stone tools found in ancient strata.

Darwin himself published in 1871 a second book called *The Descent of Man,* in which he discussed evidence showing man to have descended from subhuman forms of life. For one thing, man contains many vestigial organs. There are traces of points on the incurved flaps of the outer ear, dating back to a time when the ear was upright and pointed, and there are tiny, useless muscles still present that are designed to move those ears (some people can even today use them to wiggle their ears). There are four bones at the bottom of the spine which are the remnants of a tail, a sign that man's ancestors did have tails. In short, man and the manlike apes had a common ancestor several millions of years ago, and the entire ape and monkey tribe (the order of *primates,* from a Latin word meaning "first") had a common ancestor even longer ago.

(The antievolutionists seized upon this to declare over and over again that Darwin claimed that man had descended from monkeys, which was, of course, a distortion. No living monkey and no living species, for that matter, are ancestral to man, nor were any claimed to be by Darwin or any other reputable evolutionist.)

Scientists other than Lyell came to Darwin's side early in the game. In Germany, the biologist Ernst Heinrich Haeckel was a powerful proponent of evolution, and in the United States the botanist Asa Gray (of Harvard) carried the ball. In France, progress was slower because of the influence of Cuvier's memory (Cuvier himself had died in 1832), but even there its victory could not be long delayed. By 1880, the scientific fraternity had been mostly won over, and the doctrine of the immutability of species was just about dead.

However, the battle continued, for this was one theory in which ordinary people, who were not scientists, were also deeply involved. If Darwinism won out, what would be left of the biblical story of the Creation? The book of Genesis could be interpreted allegorically, perhaps, and made to fit Darwin, but this didn't satisfy many people who would not compromise but who insisted on a literal interpretation of every word in the Bible. (These were called "fundamentalists.") Controversy, therefore, was bitter.

The man who did more than anyone else to win the battle for evolution among educated nonscientists was the English biologist Thomas Henry Huxley. Throughout the 1850's he had been a firm believer in the immutability of species and had even argued with Darwin about it. However, when *The Origin of Species* appeared, Huxley found himself swept away by it and was at once converted from an opponent to an ardent advocate. In 1863, he, like Lyell, wrote a book on the evolution of man. It was entitled *Man's Place in Nature.* Thereafter, his writings and his lectures were read not only by scientists but also by laymen, and his views were expressed so forcefully and well that more and more were won over by him.

Beginning in 1890, tangible evidence of man's ancestors was found, for fossils of primitive men, with apelike features and small brains, were found in several parts of the world. Some argued that these were ordinary men who had suffered from diseases which had distorted their skeletons. However, anatomists could tell a distorted man's skeleton from a skeleton that was intermediate between ape and man. These fossils were what the press began to call "ape men" or "missing links."

Still the fight went on, and the last major battle occurred

in the United States. The legislature of the state of Tennessee, alarmed at the thought that children were being taught what some considered to be atheism and immorality, passed a law in 1925 forbidding any teacher in the public schools to teach that man had evolved from lower forms of life. In this they were strongly supported by the powerful fundamentalist sects.

In that same year, at a high school in Dayton, Tennessee, a young biology teacher named John T. Scopes was persuaded to tell his class about Darwinism in order to test the constitutionality of the law. Scopes did so, and in July, 1925, he was put on trial. The case (familiarly known as the Scopes trial) attracted world-wide attention.

The local population and the judge were all antievolutionary. William Jennings Bryan, a famous American politician (and probably the outstanding fundamentalist in the nation), was one of the prosecuting attorneys. To defend Scopes, a number of lawyers, including the famous Clarence Darrow, made their appearance.

The trial was very largely a farce, since the judge did not allow the defense to place scientists on the stand to testify to the evidence behind the Darwinian theory, but restricted the matter entirely to the point of whether Scopes did or did not teach the theory, a point concerning which there was no argument, after all.

The climax came, however, when Bryan, a self-styled expert on the Bible and on religion, offered to allow himself to be cross-examined by Clarence Darrow. Darrow promptly showed that Bryan was completely ignorant of modern developments in science; that he knew nothing of any religion but his own; and that his beliefs were those he had learned at his mother's knee, his "expertness" not extending an inch beyond that.

(This is not to say that beliefs learned at one's mother's knee are necessarily wrong. Still, anyone who wishes to dispute the comparative merits of belief A and belief B ought to have some knowledge of both, regardless of which is believed true and which false by the person doing the arguing. Strongly to condemn a viewpoint you know nothing about is intellectually dishonest.)

Darrow did, however, get Bryan to admit that the days of Creation were not necessarily literal days in the usual sense, but might represent eons of time. This offended other fundamentalists who thought Bryan was too radical in this.

The trial ended with Scopes being convicted and fined a hundred dollars, but the conviction was later reversed on technical grounds by the Tennessee Supreme Court. Bryan died a few days after the end of the trial.

Although theoretically Scopes had lost, most people in the United States were painfully aware that their country had been made to look ridiculous in the eyes of the educated world. The Tennessee law has been a dead letter since and there has been no further serious antievolutionary stand. Today, although many educators play it safe by calling evolutionary ideas "theory" instead of "fact," there is no reputable biologist who doubts that species, including *Homo sapiens,* have developed with time, and that they are continually, though slowly, changing.

The notion of evolution had to contend not only with the clamor of opponents, but also with the distortions of certain proponents. For instance, one of the leading evolutionists in Darwin's time was Herbert Spencer, an English philosopher. He it was, in fact, who made the word "evolution" popular. (Darwin himself rarely used the word.)

Spencer was primarily interested in the development of human societies and was the founder of the modern science of *sociology.* When Darwin's book came out, he saw at once that the notions of evolution could be applied to sociology. If species could be formed by the forces of natural selection, why not human societies as well? Thus, he founded *social evolution.*

But Spencer invented a phrase in the process that caught on at once. That was "the survival of the fittest." Others seized upon it to justify all that was evil and unpleasant in the society of the times.

Was there unrestrained competition in business with no holds barred? Why, that only led to the "survival of the fittest." Was there unemployment? The "less fit" would starve to death and the laborers who survived would be a stronger breed. Unemployment was good for them. In the same way, war weeded out the "unfit" and allowed better and stronger nations to survive. And, of course, there were also those who used evolutionary reasoning to show that one particular group of mankind (invariably the group to which the reasoner belonged) was superior to others.

The cruelties of unrestrained competition, of militarism and of racism all existed before Spencer. They were not invented by evolution. However, the last half of the nineteenth century saw people begin to justify these ancient evils by the use of "modern science." This distortion of Darwinism made the whole notion of evolution seem unpleasant to people and strengthened the hand of those who claimed evolution to be immoral and sinful.

Because there are still many people who will use the notion of "the survival of the fittest" to justify ways of life that seem to most of us to be evil, I would like to spend some time discussing the matter.

In the first place, the phrase "the survival of the fittest" is not an illuminating one. It implies that those who survive are the "fittest;" but what is meant by "fittest"? Why, those are "fittest" who survive. This is an argument in a circle.

Actually, what does one mean by the "fittest"? Suppose you were asked the question: "Which is the 'fitter,' a man or an oyster?"

Obviously a man is a much more highly organized creature, with a more efficient set of body machinery, and with tremendously greater versatility and potentialities. Who would deny that man was fitter?

But if every bit of land were suddenly placed under shallow water, then men would die and oysters would not. If mere survival is the measure of the "fittest," then under that new condition oysters would prove "fitter" than man.

In other words, "fitness" is a relative term, and has no meaning unless you mention the environmental niche you are considering. A great many species have become extinct, yet have left behind close relatives who still survive. Not only does man exist, but also rabbits, sharks, earthworms and jellyfish. The most primitive creatures who ever existed are still represented today and are even flourishing. In fact, if mere survival is the criterion of "fitness," then the king crab is many times "fitter" than man, for it has existed as a species much longer.

Of course, each species exists within its own niche, and within that niche it would have competed with other (now extinct) species. It would have showed itself "fitter" by surviving.

If we are to try to apply the notion of "fitness" to the development of human society, let us consider not just man, but also his environmental niche.

If two men and a woman were stranded on a desert island, the environmental niche would be the desert island and all it contained. If one man killed the other (by superior force or by superior guile), he would inherit the woman, so to speak, and possibly leave descendants, while the victim would not. The murderer would be "fitter" by Spencer's test, for he had survived; and by Darwin's test, too, if he left descendants.

If the same two men and a woman were in New York City, however, their environmental niche would be not only the buildings of Manhattan and the air above it and rocks below it. Their environmental niche would include all the machinery of a human society by which they would be affected as profoundly as by the inanimate environment.

This society reacts upon the individual. For instance, murderers, as a class, are a danger to society and not just to their individual victims. As long as some individuals feel free to kill, all other individuals must feel unsafe. Therefore, in all societies, even in the most primitive, murderers have, in one way or another, been hunted down and killed.

The result is that, when a human society forms part of the environmental niche, a murderer might be "fitter" than his victim, but he will be "less fit" than nonmurderers as a group, by the Spencerian test of survival.

One can also argue, in similar fashion, that within a society the dishonest businessman is "less fit" than the ethical one; that war is "less fit" than peace; that slavery is "less fit" than brotherhood.

To reach the same conclusion by another path, I might point out that "competition" must be understood in a broader sense than that of a fist fight. Competition among the individuals of a species may be a competition of comparative cooperations. One factor in the survival of a species has often been its ability to live in packs; to have one individual of a pack act as watchman while the rest graze; to be in the habit of defending themselves as a pack against an attacking enemy when individually no defense might be possible. (Predators also can hunt in packs and hunt more successfully than if each animal went off on its own.)

Any improvement in the "pack habit" increases the chances of that species' survival. Moreover, if there is variation among the species so that some groups have more of the "pack habit" than others, it is those with the better "pack habit" that will survive.

The same is true within the human species. History is full of examples of peoples who, unable to cooperate among themselves, fell under the onslaughts of others, who were individually perhaps less admirable and advanced, but who had the virtue of cooperative action. The fate of the ancient Greeks is the most tragic case in point. By temporarily uniting, they beat off the Persians; by being unable to unite later, they fell to the Macedonians.

Any society which indulges too extensively in the Spencerian notions of "the survival of the fittest" will break down through internal dissension and fall prey to other societies which are less Spencerian. We have now reached the point where it seems that unrestrained competition among nations may do us all in as a species and that some sort of non-Spencerian cooperation is essential or we will have finally proved our "unfitness" by not surviving.

If Darwin's theory managed to survive opposition and distortion, that did not mean it was flawless. As a matter of fact, it had a serious weak point, which Darwin himself recognized.

As mentioned before, Darwin had no real explanation for variation in physical characteristics. It happened, that was certain, but why?

Worse still, Darwin thought that variations consisted of infinitely small differences and that when two parents varied in some respect, the youngsters were intermediate in that same respect. But if that were so, then when mating occurred generation after generation, should not the variations average out? Should not intermediacy become universal?

To put it as simply as possible, how did variation come about in the first place, and what made variation persist long enough for natural selection to get in its work?

Some of the Darwinians felt the lack and tried to supply reasons. Several thought that variations did not proceed by infinitesimal steps. They suggested that evolution proceeded by jumps, so to speak. Every once in a while there might be a large variation, one too large to average out before natural selection had established it.

Although this seems a daring speculation, there was actually considerable evidence in favor of exactly this happening. Over and over again herdsmen and farmers noticed the birth of strange varieties among their livestock and crops. These anomalies were viewed with suspicion and mistrust and were generally considered warnings of divine displeasure. (In fact, the strange varieties are often called "monsters," from a Latin word meaning "to warn." A less emotional term is "sport.")

It was not until comparatively modern times that superstition and general uneasiness gave way to the thought that sports could be made useful. In 1791, a male lamb belonging to the flock of Seth Wright, a Massachusetts farmer, was born with unusually short legs. When it grew to maturity, it was bred to ewes, and the lambs that resulted were likewise short-legged. Eventually, a whole herd of short-legged sheep resulted. The advantage of the short legs was simply this: the sheep could not jump over the low fences surrounding the pasture and were, therefore, less troublesome to keep. This early breed appeared again. this time in Norway, and the short-legged breed was reestablished.

Since 1791 many other useful sports have been discovered and bred. It seems certain, furthermore, that long before 1791, even back in prehistoric times, sports must have been preserved and bred, and this accounts for the numerous breeds of dogs and other domestic animals that have existed through the centuries.

Yet all this material in connection with sports took considerable time to penetrate science itself. After all, scientists knew little about the mechanics of herding animals and cultivating plants, while herdsmen and farmers, for their part, did not write papers describing their discoveries.

So it was not until 1884 that a book was written which systematically presented evolution as occurring by jumps. This book was by a Swiss botanist named Karl Wilhelm Von Nägeli. Even the existence of jumps did not seem enough to account for all the facts. Why did the jumps not average out? So Nägeli went on to suggest that there was some drive within the species that kept it varying in the same direction. Once a species started jumping, say, in the direction of increased size, it continued jumping in that direction faster than ordinary mating could level out the size again. In this way, the species would grow larger and larger, as the primitive horses of past ages grew from the size of a dog to the giant animals of today. A species might even grow larger than was desirable (as if it were going too fast to stop) and might become extinct through "overlargeness." This kind of "biological inertia" was called *orthogenesis*.

Though Nägeli's theory of orthogenesis was not accepted, his notion of discontinuous evolution persisted. A Dutch botanist, Hugo De Vries, set out to find as much evidence as he could for the actual occurrence of sudden large variations in species.

In 1886, De Vries came across a wild colony of the American evening primrose in which some of the individual plants were quite different from the rest. If they were crossed, they produced a new generation like themselves and not like the ordinary primrose. With continued investigations, he found new sudden changes. He called these *mutations,* from a Latin word meaning "to change."

His experiments in crossing plants also taught him a few things about the manner in which physical characteristics are inherited. By 1900, he had enough experimental evidence to feel himself ready to publish a complete theory on inheritance.

Although De Vries was not aware of it, two other botanists were making much the same observations as he was, and were getting ready to publish essentially the same theory. These were the Austrian, Erich Tschermak, and the German, Carl Erich Correns.

All three, working independently, went through previously published material on the subject, once they had worked out their theories. And all three found they should have done this first, because all three, looking through an obscure journal, *The Proceedings of the Natural History Society of Brünn,* found a paper by someone they had never heard of, someone with no scientific reputation, someone who was merely an amateur gardener.

That paper, however, that piece of work by an amateur gardener (an Augustinian monk named Gregor Mendel), detailed in full the theory which De Vries, Tschermak and Correns had each worked out independently. What is more, that original paper had appeared in 1866, thirty-four years earlier.

Each of the three scientists was true to the ideals of science. Each called the attention of the world to Mendel's paper. Each gave Mendel all the credit, and to this day the rules that govern the inheritance of physical characteristics are referred to as Mendel's Laws.

Here is the story of Mendel, as rediscovered by the three botanists.

During the 1860's, Mendel taught natural history in the monastery school at Brünn (now Brno, Czechoslovakia) and also tended the garden. He amused himself by carefully crossing plants and observing the exact results.

He worked with pea plants which existed in his garden in a number of sharply marked-off varieties, which, however, were all the same species since they could all be "crossed."

For instance, there was a variety of pea plant with a purplish-red flower, and a variety with a white flower. Peas of the red variety, crossed among themselves, produced seeds that grew into plants with red flowers only. Those of the white variety, crossed among themselves, produced seed that grew into plants with white flowers only.

But what if the red variety were crossed with the white variety? Mendel did this and observed that all the seeds developed in this cross grew into plants with red flowers. Not one white flower in the bunch! The white flower characteristic had disappeared!

Or had it? Mendel next crossed the peas of this new generation of red flowers among themselves, waited for seeds and planted them. Then, in the third generation, some of the seeds grew into plants with white flowers, pure white. To be sure, these were in the minority. Of all the third-generation seeds, almost exactly one-quarter gave plants with white flowers. The rest grew into plants with red flowers.

Next he tried another generation. Suppose the whites of

the third generation were fertilized among themselves or, better yet, were fertilized with their own pollen (self-pollination)? Only white-flowered pea plants resulted.

Suppose the third-generation red-flowered plants were self-pollinated? Here two different results showed up. There were some reds that produced only red-flowered offspring. There were others that produced both red-flowered and white-flowered ones in the ratio of three to one.

In other words, white-flowered pea plants always "bred true." Red-flowered pea plants sometimes bred true, and sometimes did not.

To explain the results of his experiments, Mendel devised a theory. He supposed that each plant contained two factors which controlled a particular characteristic, such as the color of its flowers. Today we call such factors *genes*, from a Greek word meaning "to produce."

Thus, a plant with red flowers might have two genes, each tending to produce red flowers. If we symbolize such a gene as *R* (for "red"), then the red-flowered plant containing two such genes, could be called an *RR* plant. Similarly, a white-flowered plant would contain two genes tending to produce white flowers and could be called a *WW* plant.

Mendel next supposed that each plant transmitted only one gene apiece to the ovule or to the pollen grain. The combination of the two in the process of *pollination* gave the offspring a total of two genes again.

The *RR* plant could only transmit an *R* gene to either ovule or pollen, so that when an *RR* plant is self-pollinated or cross-pollinated with another *RR* plant, only *RR* offspring result. Similarly, *WW* plants can produce only *WW* offspring.

If an *RR* plant is crossed with a *WW* plant, however, the pollen of the first carries only an *R* gene, the ovule of the latter a *W* gene and the offspring have one of each. The second generation of such a cross consists exclusively of *RW* plants. (If it is the pollen of the *WW* plant that pollinates the ovule of an *RR* plant, the result is the same. The combination of *W* and *R* still gives an *RW* plant.)

But all such *RW* plants produce red flowers only. Apparently, the presence of the *R* gene drowns out the presence of the *W* gene.

Nowadays, we call two genes which both govern the same characteristic in different ways *alleles*. Thus the R gene and the W gene are alleles because both govern flower color. Since the R gene produces its effect even in the presence of a W gene, the R gene is said to be the *dominant* allele, while the W gene is *recessive*.

However, what happens when an RW plant is self-pollinated? It can pass on only one gene to its pollen grains and it shows no bias in favor of either. Half the pollen grains are R and half are W. In the same way, half the ovules are R and half are W. In the pollination process, the resultant offspring can be produced in the following manner: (1) by the pollination of an R ovule by an R pollen grain to produce an RR individual; (2) by the pollination of an R ovule by a W pollen grain to produce an RW individual; (3) by the pollination of a W ovule by an R pollen grain to produce an RW individual (RW and WR prove to be the same); (4) by the pollination of a W ovule by a W pollen grain to produce a WW individual.

All four alternatives are equally probable and happen in roughly equal numbers. Three-quarters of the plants (RR, RW and RW by alternatives 1, 2 and 3) have red flowers. The remaining quarter (which were the WW plants produced by alternative 4) have white flowers.

Suppose, next, that a hybrid red-flowered plant, RW, is crossed with a white-flowered plant, WW. The hybrid plant would produce R pollen grains and W pollen grains in equal proportion and R ovules and W ovules in equal proportion. The white-flowered plants would produce only W pollen grains and W ovules. The only two possibilities when crossed would be that (1) a W ovule combined with a W pollen grain or (2) a W ovule combined with an R pollen grain. This means that if the pollen of the RW plant were used, the offspring would be half WW and half RW.

If the pollen of the WW plant were used, then there would be two possibilities again: either (1) a W ovule combined with a W pollen grain, or (2) an R ovule combined with a W pollen grain. Here, too, the offspring would be half WW and half RW.

In either case, Mendel's theory would predict that half the plants that resulted from the cross would produce red flowers and half would produce white flowers. When tried experimentally, this turned out to be so.

Mendel did not test flower color only in crossing his pea plants. Actually, he chose seven different characteristics that varied from plant to plant. There were plants with yellow seeds and others with green seeds; plants with round seeds and others with wrinkled seeds; plants with tall stems and plants with short ones and so on.

In each case, when he crossed one variety with another, one turned out to be dominant. Round seeds were dominant over wrinkled seeds, yellow seeds were dominant over green seeds, and so on. The hybrids all produced a third generation in which the recessive form showed up again in a quarter of the total.

Furthermore, each of the seven characteristics was inherited independently. For instance, a particular plant might inherit red flowers and long stems, red flowers and short stems, white flowers and long stems, or white flowers and short stems. And any of these combinations might go along with yellow, wrinkled seeds or green, wrinkled seeds; yellow, smooth seeds or green, smooth seeds. All possible combinations of the seven characteristics could develop, so that by proper crossing one could end with 128 different varieties of pea plants.

Mendel's results explained some of the difficulties that Darwin had run into. For one thing, variations among offspring did not run along an entire gamut of infinitely small steps. There were large, discrete differences. An individual plant might have red flowers or white flowers; there need be no in between.

Secondly, there was no "blending" of inheritance. Crossing red-flowered plants and white-flowered plants resulted in red-flowered plants and not varieties of pink. What is more, even when a recessive gene seemed lost and a particular characteristic seemed to have disappeared, it was still there and would appear, unharmed and unchanged, in a later generation. Let two W genes come together, no matter how many generations had elapsed during which they were in the constant, overpowering presence of R genes; let them but come together and a white-flowered plant is the result.

The objections of Darwin's opponents that, with random mating, variations would average out and yield one long, dull mediocrity were, therefore, not valid.

Mendel wrote up his observations and sent them to Nägeli. Nägeli was not impressed, because he thought Mendel was just blindly counting plants instead of working up some fundamental new scheme like his own orthogenesis.

This was a bad break, for Mendel's theory was of fundamental importance, while Nägeli's theory was worthless. However, it was Nägeli who had the reputation and not Mendel, so the poor monk published his paper in the obscure journal and did not try to follow it up. His work remained unknown, and he himself unhonored.

Darwin died in 1882, never knowing that the greatest weakness in his theory had been patched up. Mendel died in 1884, never suspecting that he was destined for fame. Nägeli died in 1891, never dreaming what a terrible mistake he had made.

5

Progress at Random

The gene theory of inheritance is not as simple, all told, as it would seem from Mendel's experiments on the seven variations in the pea plant, and the science of inheritance (or *genetics*, as it is called) has become a complicated one, indeed.

For instance, genes cannot always be clearly differentiated into dominant and recessive. As an example, consider the human blood types. The most common types, those which are important in transfusions, are under the control of three alleles, which can be symbolized as an *A* gene, a *B* gene and an *O* gene.

A human being with a pair of *A* genes (an *AA* individual) belongs to blood type *A*. Similarly, a *BB* individual would belong to blood type *B*, and an *OO* individual to blood type *O*. If an *AA* mates with an *AA* (or a *BB* with a *BB*, or an *OO* with an *OO*), all the children are *AA* (or *BB* or *OO*) and are of blood type *A* (or *B* or *O*).

But suppose there is crossmating. Suppose an *AA* individual mates with an *OO* individual. The *AA* contributes one gene to the offspring and this one gene can only be an *A*. The *OO* contributes one gene which can only be an *O*. The offspring, combining the two genes, is an *AO* individual. If his blood is tested, it proves to be of blood type *A*. Apparently the presence of the *A* gene completely masks the presence of the *O*. The *A* gene is dominant over the *O* gene.

Exactly the same thing happens in a mating between a *BB* individual and an *OO* individual. The offspring are all *BO*, and the blood type is always *B*. The *B* gene is dominant over the *O* gene. So far, the situation is just the same as with Mendel's pea plants. (Part of the value of

Mendel's law is its perfect generality. It applies to all species, from pea plants to humans.)

Furthermore, if an *AO* individual is mated with another *AO* individual, each may contribute an *O* gene to a particular offspring. An *OO* individual of blood type *O* would have arisen, so that that recessive characteristic will have reappeared unharmed after having skipped a generation. The mating of a *BO* with a *BO* produces similar consequences. Again, there is no difference here between men and pea plants.

Now, however, suppose an *AA* individual is mated with a *BB* individual. The *AA* individual contributes an *A* gene, the *BB* individual a *B* gene. All the offspring are, therefore, *AB*. But what of the blood type? Is *A* dominant over *B* or *B* over *A*? The blood, when tested, shows that neither is dominant. The blood reacts as both type *A* and type *B*, and such a case is classified as blood type *AB*. Each gene shows its full effect despite the presence of the other.

Incomplete dominance need not result in a display of the effects produced by both genes; it can show an effect produced by neither. There are plants that exist in red-flowered varieties and white-flowered varieties which, when the two varieties are crossed, produce plants with pink flowers.

The red and white colors seem to have blended, but they have not. If pink-flowered plants are crossed with pink-flowered plants (or if they are self-pollinated), the red-flowered genes and white-flowered genes sort out in all possible combinations. The offspring produce red flowers (*RR*) in one quarter of the cases, white flowers (*WW*) in another quarter, and pink flowers (*RW* or *WR*) in the remaining half.

Genes possessing two alleles (red flowers and white flowers or smooth seeds and wrinkled seeds) or even three alleles (*A, B* and *O* blood types) are relatively easy to work with. However, genes may possess large numbers of alleles. The gene governing the *Rh* blood types in man possesses at least eight alleles of varying types of dominance with respect to each other. The inheritance of the *Rh* genes is, therefore, so complicated that controversy concerning it continues and will probably go on for quite a while yet.

Moreover, a particular physical characteristic may be governed by the interaction of more than one set of alleles, and this further complicates the problems of inheritance. Human skin color is probably a case in point here. It is not even agreed as yet how many different sets of alleles there are contributing to this physical characteristic. The presence of more than one set makes for a large number of gradations between the extremes (and we know from experience that there are a great many color levels in the skin intermediate between that of the Swede and that of the Nigerian). It is the complicated gene pattern of such easily noticeable physical characteristics that gives the casual observer (and even one not so casual, such as Charles Darwin) the illusion that the characteristics of parents mix and blend in the offspring.

Sometimes the effect of a gene will be altered by circumstance. For instance, there is a gene, or possibly a combination of genes, that produces early baldness in a human being. There is reason to think, however, that if the circulating male sex hormone is below a certain level of concentration in the blood, baldness does not develop even in the presence of the baldness gene or genes.

If the level of circulating male sex hormones is controlled by another gene (or genes), as it well may be, this would be the case of one gene affecting the action of another.

Then, too, the action of a gene may be affected by the environment itself. A person may possess both the gene for baldness and the gene governing a high concentration of male sex hormone. Ordinarily, this would ensure early baldness. If he is castrated in childhood, however (a purely environmental effect), the male sex hormone supply is cut off at the source, and he will not go bald. (This is a baldness preventative that is unlikely to attain great popularity.)

Similarly, a person with a gene, or combination of genes, which would produce, under ideal circumstances, a height of better than six feet, will, nevertheless, not attain that height if he has been chronically undernourished as a child.

The pattern of inheritance of physical characteristics among the higher animals is studded with pitfalls like this, and it is no accident that most of the advances in genetics have been made as a result of experiments on plants, bacteria and insects. There we have the happy combination of many offspring in a short time so that statistical rules will apply well, plus the existence of relatively few and uncomplicated physical characteristics.

Human genetics is a particularly poorly studied field. Not only does the individual human being have few children only at long intervals, but, worse still, controlled matings are difficult or impossible to arrange among human beings. The human being is one of the poorest laboratory animals in existence, and the scientist studying human genetics must take what he can get in the way of mating combinations.

Furthermore, the human being possesses characteristics in which we are all terribly interested, which are not only particularly subtle but the study of which is impossible in other animals. About such characteristics as creative genius, musical talent, chess mastery, or fertility of imagination, virtually nothing is known from the standpoint of genetics.

It is this sort of thing that has dashed the hopes of the early evolutionists that perhaps, once the mechanism of evolution was understood, the future evolutionary course of humanity could be controlled. Undesirable traits, they had hoped, might be eliminated and desirable ones stressed so that evolutionary progress could be speeded up.

The pioneer in this belief was Francis Galton, a cousin of Charles Darwin. He studied, with statistical thoroughness, the appearance of certain traits in family lines and in 1883 wrote a book in which he discussed methods for improving the human species through the deliberate encouragement of proper matings. He referred to this as *eugenics*, from a Greek word meaning "well-born" (and not to be confused with "genetics"). Galton's work, be it noticed, appeared after the gene theory had been developed, but before anyone but Mendel knew about it.

The end in view is a noble one, but eugenics has foundered on the rock of ignorance and impracticability. We just do not know enough about human heredity to plot intelligent marriages. And if we did, voluntarily controlled matings would be difficult to arrange, since it is unlikely that human beings will abandon their habits of marriage according to taste and adultery according to opportunity. Controlling matings by force, as well as by use of sterilization, would be violently repugnant to most people. (Who would feel safe?)

In connection with our ignorance of human genetics, it is pointed out sometimes that men, by controlling the matings of domestic animals, have "improved the breed." This was done despite the fact that the genetics of horses and cattle is not much better known than human genetics. In fact, most of the improvement of the breeds took place in eras when there was not the foggiest notion that the science of genetics existed or could exist.

If this is so, could we not improve the breed of men, working equally intuitively?

This argument sounds good and was used as long ago as 700 B.C. by the ancient Spartans. They condoned and even approved of adultery when the "other man" in question had desirable characteristics, and pointed out in explanation that human breeding should be given at least the same care as the breeding of cattle, and that a desirable male in either case should have all the offspring possible.

And yet such an argument is worthless. We have clear notions of what constitutes an "improvement of breed" in domestic animals. If we want a cow that is a good milker, we interbreed bulls and cows that have descended from good milkers, and pick the best of the offspring (in that one respect) for future interbreeding. In the end, we build up milk-specialists that are scarcely anything more than living factories, designed to turn grass into butterfat.

Fine! But what else have we bred into the cattle while we have been concentrating on the milk? We do not much care; we just want the milk. Our tame cattle are now too stupid and placid to protect their calves or even themselves against wild beasts. The thoroughbred race horse is a magnificent speed-machine, but is a highly neurotic creature that requires more and better care than a human baby. Swine have been made into ungainly lard-producing creatures fit for nothing but eating and being eaten; dogs have been made insanely combative (as in the case o the Pekingese) or so snub-nosed they can barely breathe (as in the case of the English bulldog), and so on.

It is all very well to breed domestic animals for a particular characteristic at the expense of their over-all survival, since we are taking care of them and, in any case, want them to meet our needs and not their own.

In the case of *Homo sapiens*, however, for what do we breed? The ancient Spartans thought they knew. They were interested in the various warlike qualities such as strength, endurance and courage. This they accomplished; the behavior of the Spartans in battle at Thermopylae stirs our admiration (though few of us would care to emulate them). However, the neglect of all other characteristics produced a Spartan culture that is completely unadmirable and, in fact, is the clearest example of a long-lasting psychotic culture available in history.

Again, in the case of *Homo sapiens,* for what do we breed? For athletic excellence? For martial virtues? For long life, creative genius, sunny disposition, literary ability, mental stability, cool wisdom, a strong sense of ethics? All are desirable, in their way, but there is no way known to breed for all at once; and experience with domestic animals, or with Spartans, gives us no hint of how to breed for numerous characteristics at the same time.

There are, of course, extremely bad characteristics (such as idiocy or homicidal mania) which we would like to breed out, if we knew how. Yet we are not entirely certain that we can get rid of even undesirable genes without also getting rid of a certain proportion of desirable ones. To put it as simply as possible, there have been many great men, for whose existence the world is most grateful, who have been epileptics, diabetics, homosexuals or serious neurotics.

The result is that reputable geneticists approach the whole question of eugenics with the greatest caution, and those who are most enthusiastic for eugenics tend to be cultists who are filled with confidence through ignorance and who tend to veer from eugenics into racism.

A particularly important factor introduced by the gene theory, as far as evolution is concerned, is that of randomness. No matter how complicated the study of genes and their interrelations gets, one thing always remains. Individual genes sort themselves out from generation to generation in a completely random fashion.

Yet before the gene theory was established, there was considerable difficulty in believing that the development of species could be entirely a random affair. Could an ameba become a man through the operation of blind and random forces? It seemed there had to be some sort of purpose behind it. One way or another, there had to be direction. For instance, Nägeli's orthogenesis was a kind of purposefulness superimposed upon evolution. Species did not grow bigger or smaller at random, according to orthogenesis; once they started growing bigger, they continued willy-nilly. In the same way, an ameba that started out in the direction of the far-distant goal of man would tend to keep on the road and not stray from it.

It may well have been the implication of randomness in Mendel's theories that set Nägeli's teeth on edge and caused him to pooh-pooh Mendel's paper.

And yet to account for the facts of evolution, randomness of gene shuffling as proposed by Mendel is all that is needed, provided it is combined with the natural selection postulated by Darwin, the mutation theory suggested by De Vries and the eons of time put forward by Hutton. Using all these, we can build up a simplified scheme of the evolution of the large cats—simply as an idealized example.

Imagine members of a species of large cat living throughout Africa, southern Asia and northern South America. We can suppose them all to have identical genes and each gene to be made up of a single allele. (Such an ideal case would never actually be met with in nature.)

In accordance with the De Vries mutation theory, as genes are passed on from generation to generation, new alleles are developed. Eventually, every gene would be made up of a number of such alleles. According to Mendel's laws, these alleles would be shuffled into new and random combinations in each generation. (In fact, the evolutionary significance of sex is that it supplies a method for the shuffling of genes by having offspring collect half their genes from each of two parents. At which thought one might cry, *"Vive la signification évolutionnaire!"*)

No two cats are very likely to develop the same new

allele, but that makes no difference. Through unrestricted mating, an allele developed by one cat spreads gradually through the entire species. If, however, certain portions of the species are geographically isolated, the alleles developed in one portion will not spread to the others. Thus, there will be special alleles developed by the Asian cats which neither African nor South American cats will ever receive (unless they produce them independently, as, by the operation of chance, they may). The same argument holds true for special alleles developed by the African or South American cats.

The three groups will, with time, begin to show consistent differences in physical appearance, and these differences will increase with further time.

The new alleles, as they are formed, will be subject to the forces of natural selection. An allele producing a characteristic that is somewhat undesirable will be discriminated against slowly. Eventually, it will die out, except that it may be regenerated every once in a while by a new mutation. The tendency to die out will be balanced by the tendency for regeneration so that the allele will persist at a certain equilibrium level. The more desirable the gene, the faster it will be discriminated against, and the lower the equilibrium level.

On the other hand, an allele that produces a desirable characteristic is naturally selected, since cats possessing it will live longer and breed more. If the advantage is sufficient, that allele may replace all others.

The comparative worth of a particular allele depends upon the environment to which the animal is subjected, and one that is beneficial in one environment may be detrimental in another. The differences among the Asian, African and South American cats which are initiated by chance mutations would then be accentuated by the action of natural selection (which is not random by any means).

Thus, if the African cats are creatures of the plains, a tawny coat, produced by a particular line of mutations, may be of particular use to them. Such a coat would render them unnoticeable against the background and increase the efficiency with which they can stalk game. (On the other hand, if it had a striped or spotted coat pattern, it could be seen by a half-asleep giraffe at a distance of two miles, and the cat would probably starve to death.) The tawny coat produced slowly by the random action of mutations and gene assortment would then be established as a universal characteristic among the African cats by the comparatively rapid action of natural selection.

The Asian and South American cats, however, might live in a jungle where the general background is one of sunlight filtering through leaves to make a splotched pattern of light and dark. There, a striped or spotted coat pattern would render them particularly unnoticeable. The Asian cats might develop, by random mutation, a striped coat, and natural selection would fix that as a universal characteristic. The South American cat might develop, by equally random mutation, a spotted coat, and natural selection would fix that. Had the South American cats developed stripes and the Asian cats spots, matters might have been fixed that way just as easily, but that alternative did not happen, for no better reason than that which causes a coin to fall heads, and not tails, at some particular throw, or for a pair of dice to turn up snake-eyes instead of boxcars, or a poker dealer to turn up the ace of hearts instead of the jack of diamonds.

In this imaginary development of lions, tigers and jaguars, we have the germ of a treatment for all evolution.

The over-all picture is that of groups of organisms which, by mutation, random assortment, and natural selection, and without need of any directing force at all, continually improve the manner in which each hits the particular environmental niche in which it lives. Occasionally a mutation, or series of mutations, enables a species or a group of species to progress from one evolutionary niche into another (by stages during which it lives in intermediate niches, perhaps). This new niche may be empty or it may be filled with creatures less well adapted to it than are the newcomers.

The extent to which an organism fits its particular environmental niche is the measure of its *specialization*. The greater its specialization, the better it fits the niche, the more efficiently it exploits it, and the more secure it is in its position, provided the niche remains stable.

However, the more it increases its specializaton with respect to a particular niche, the less a particular organism is able to live outside its niche. Anything which wipes out its niche (such as climatic changes, a fire that destroys a forest, or a blight that destroys a particular type of plant) would wipe out creatures that are so specially adapted for a particular climate, a particular forest, or a particular type of plant as food supply, that they cannot adjust to the absence of the climate, forest or plant.

Less specialized (or more generalized, if you prefer) organisms, while rather inefficient in a particular niche, are not as subject to the vicissitudes of fortune. In the case of the disappearance of that niche, they can go and be inefficient in another niche. This often works to the long-term advantage of the organism concerned. It is worth mentioning that mammals developed from a rather unspecialized group of reptiles; and that primates developed from a rather unspecialized group of mammals; and that *Homo sapiens*, except for his brain, remains one of the more unspecialized primates.

Very often, an increase in specialization goes along with an increase in body complexity. This happens often enough so that there is a popular illusion to the effect that evolution always progresses from simple animals to complicated ones.

This is not so. A better fit to a particular environmental niche may require increased simplicity. It seems to be generally true that a bodily feature for which there is no longer any use under new conditions or in a new environmental niche tends to drop out. There seems to be increased efficiency and a better chance of survival for those animals that do not bear the unnecessary load of useless features.

For instance, manlike creatures have been cooking food for perhaps a hundred thousand years, softening it, in this manner, for easier chewing. They have, for shorter periods of time, been using knives and other tools to hack it into pieces. Teeth are consequently no longer as important as they once were. The canines, originally designed to seize and tear, the incisors to slash and cut, the molars to grind, all find their jobs growing less important. Over the ages, then, developing man seems to have been gradually decreasing the strength of his jaws and the size of his teeth. The four wisdom teeth (the hindmost molars) are now more trouble than they are worth. They fit badly into the jaw and decay easily. Occasionally people are born without the capacity to form one or more of them. Obviously, human evolution is progressing in the direction of the loss of wisdom teeth and the replacement of our complement of thirty-two teeth by one of only twenty-eight—a simplification.

Those species of fish and lizards that have managed to survive by taking to cave life as an environmental niche offer another example. (Cave life is a poor and undesirable niche, but for that very reason offers the compensation of decreased competition.) Life in the perpetual darkness of a cave makes eyes unnecessary, and many cave species have weak eyes or are altogether blind. Many also lack color, since pigmentation offers no particular advantage in eternal darkness. Related species that live in the outside world and from whose ancestors the cave species have descended are fully pigmented and have perfectly good eyes.

The most extreme case of specialization by simplification involves those species of creatures which have taken, as their environmental niche, the living bodies of other species. Organisms living in other organisms are *parasites*. A parasitic life has its advantages. A parasite lives surrounded by its food and need never search for it. It can, and often does, abandon the complex paraphernalia required by its free-living ancestors in a life dedicated to the detection, pursuit and digestion of food. Nor do parasites need the devices developed to combat or escape from enemies.

A creature like the tapeworm, growing in the intestines of beasts and humans, dispenses with everything but tiny hooks up front with which to catch hold of the intestinal lining, a body surface through which it can absorb predigested food, and a reproductive system in full complexity by means of which it can produce eggs and eggs and eggs. The specialization and simplification are both extreme.

By 1900, the fusion of the ideas of Hutton, Darwin, Mendel and De Vries had supplied a united mechanism that could somehow explain almost any question involving the origin of species and the development of life from microscopic blob to triumphant man.

There was one major catch, however, and that was that the genes were philosophic concepts that were discussed but had never been observed. Was there any way of studying the machinery of inheritance by the senses rather than by lofty deduction?

As a matter of fact, by 1900 this had already been done, without the doers being aware of it. Through all the century between Hutton and De Vries, a completely different line of research had been probing into the microscopic structure of living organisms and had been coming up with facts, independently derived at, which just fit the generalized evolutionary theory of 1900. And in 1902, the two lines of theories and observations fused.

To describe how this happened, we must go back, begin at the beginning, and work through the second line of research step by step.

PART TWO

* * * *

6

Up to Chordata

During the decades when theories of evolution were advanced and defended, one counterattack might have been made which was, however, not made. That counterattack might have been to the effect that there was more than one form of life; that the differences between the forms were fundamental; and that evolution involved changes from one form to another which could not be.

There was, after all, precedent for this kind of argument, a precedent from the world of inanimate nature.

The alchemists of the Middle Ages were convinced that any mineral substance could be converted into any other mineral substance if only the proper methods were used. In fact, some even thought there was a kind of mineral evolution always going on in the soil, a process by which non-metals slowly turned into the more valuable metals, and then ascended the scale of excellence from lead through iron, copper, and silver until the perfect metal, gold, was formed. It seemed improbable, some alchemists argued, that Nature would form a substance as perfect as gold all at once, any more than it formed a tree all at once. Just as a tree grew from a seed, gold would grow from simpler, less perfect metals.

Over the centuries, alchemists devised all sorts of schemes to imitate what they thought were natural processes and to form gold out of less perfect (and cheaper) metals. All schemes to do this failed.

Finally, in the seventeenth and eighteenth centuries,

chemists slowly came to the conclusion that each metal was irrevocably and eternally itself; that only gold could be gold; that all the gold that existed was gold from he beginning and would be gold to the end. In fact, all substances were made up of elements and combinations of elements, and though the combinations might change, the elements themselves could not. The elements were permanent and immutable. (Actually, we have now come to modify that view, but only under very special conditions.)

Why could not these truths about the elements, which by the early nineteenth century were well established, be used by analogy to refute those who thought species traveled up the scale of excellence, as alchemists had thought earlier of metals? Why not point out that the species were the elements of life, as immutable as the chemists believed the elements of the mineral world to be?

The reason why this argument was not advanced was that, in the decades before Darwin's book, great discoveries were made about the structure of living tissue which demolished the argument in advance.

It began in 1665, when the English scientist Robert Hooke made a more or less chance observation. He looked at a thin slice of cork under a microscope and noticed that it was made up of tiny holes marked off by walls, something like a honeycomb. He called the little holes *cells,* from a Latin word meaning "a small room."

During the next century and a half, other microscopists observed the structure of other tissues, living or recently dead. Blood, for instance, contained little bodies that were first called "globules." Other tissues seemed to be divided into tiny units marked off by walls or, at the very least, by thin membranes.

Hooke's cork contained empty hollows that were truly cells in the literal meaning of the word, but the tiny units in most tissue seemed filled with fluid. Nevertheless, although they were called a variety of names by various people at first, in the end all the units, full or empty, were called cells. The study of these cells was termed *cytology;* the prefix *"cyto-"* comes from the Greek word for "a small room."

In 1838, the German botanist Matthias Jakob Schleiden, having found cells wherever he looked, suggested that all plants were only conglomerations of cells and that the cell was the unit of life. Another German naturalist, Theodor Schwann (whom we met in Chapter 1 as one of those who helped disprove the doctrine of spontaneous generation), extended this notion to animals as well in 1839. Animals, too, he considered to be a collection of cells.

Schleiden and Schwann together are considered the founders of the *cell theory;* that is, that all living matter consists of cells.

The fact that the cell was the unit of life could be best established if forms of life could be found that consisted of but a single cell. It was suspected that this was true of protozoa and various other microscopic creatures. The German zoologist Karl T. E. von Siebold first presented conclusive evidence in favor of this point of view.

Meanwhile, it had also been established that large organisms, composed of vast numbers of cells, began life as single cells. Observations were made of how single cells grew and eventually divided into two cells, which might separate or might remain together. Large organisms grew from a single cell by division and redivision, repeated over and over, growing at once in size and in the number of cells contained.

Finally, in 1860, the German pathologist Rudolf Virchow put it all in a nutshell with the statement (in Latin): "All cells arise from cells." Pasteur was about to prove that all life came from life (as described in Chapter 1), and Darwin had just published *The Origin of Species.* This was a heroic time in biology.

And yet it might still seem that antievolutionists could accept the cell theory and still maintain that the original cells from which one species developed were fundamentally different from those from which another species developed and that the gap was as unbridgeable as that between the chemical elements.

The complete answer to that argument came from the researches of chemists themselves, the details of which will be left for the third section of the book. Here only a few facts will be stated.

The content of living cells (as opposed to the dead and empty cells of cork) was a viscous, jelly-like substance which was called a variety of names by the early investigators. In 1839, the Czech physiologist Jan E. Purkinje suggested that the material making up young animal embryos be called *protoplasm* (from Greek words meaning "first form"). In the next year the German botanist Hugo von Mohl applied the term to contents of the cell. After all, the very earliest form of an organism is as a single cell and its contents are the true "first form."

The physical characteristics of protoplasm seemed the same through all forms of life, plant and animal, from simplest to most complex. As its chemical nature was studied, it soon became obvious that all protoplasm was alike chemically as well. One of the important contributors to that point of view, in the late 1840's, was Nägeli, who was later to be the inventor of orthogenesis and the squelcher of Mendel.

And so, by the time Darwin's book came out, it was generally accepted that, physically and chemically, life was a unit, and that the various species, however different they might seem in appearance, were but variations on a single theme. The argument from the chemical elements could not be, and was not, made.

The existence of the cell theory extended the evolutionary background of life beyond the record of the fossils. Even the earliest fossils, dating back some half a billion years, are already quite complex in comparison with some of the simpler organisms existing today; and stretching beyond and back of the earliest record there must have been a time when all life started as individual cells, just as all individual organisms do.

By the time the first fossils made their appearance in the Cambrian era, all the phyla but one (the exception being the phylum to which we belong) had already been developed; our own appeared in recognizable traces not long after.

Obviously, there must have been a long history of life in the pre-Cambrian eras, the time before fossilization, during which the various phyla developed out of ancestral forms.

A connected picture of the gradual development of man from the earliest free-living cells in the primordial ocean can be built up from the fossil record, from the structure of animals now living, and from judicious guessing.

We can begin by picturing a primordial ocean existing many eons before the oldest fossil-containing rocks were formed. This ocean, for reasons discussed later in the book, probably contained considerable concentrations of substances that could serve as food for primitive cells. Living on this food, growing and multiplying, were such primitive cells.

One-celled organisms possess genes as we do; produce new alleles by mutation; divide into two cells periodically, each daughter cell being presented with the new combination of alleles. What is more, the cells are subjected to natural selection in a way that varies from region to region of the ocean. The different sections of the ocean vary in temperature, light intensity, water pressure, mineral concentration and so on, from area to area and from depth to depth.

All the evolutionary pressures are present, and in the course of hundreds of millions of years (there being plenty of time in the earth's long term of existence) the one-celled organism would develop into numerous varieties, into species, genera, orders and classes.

Such is the successful adaptation of the one-celled organism generally to life on earth that many forms have survived to the present day essentially unchanged except, perhaps, for increasingly specialized adaptation to their environmental niches.

At some early date, one cell or group of cells must have developed a compound called *chlorophyll*. This freed them from the necessity of depending upon the food substance in the ocean. By the use of chlorophyll, sunlight could be absorbed by the cell and its energy used to convert water and carbon dioxide (a gas present in the air) into foodstuffs which it could store and use at leisure.

This involved such a fundamental change in the nature of the cell machinery that it served as the basis for the division of all life forms into two kingdoms. Cells containing chlorophyll belonged to the *plant kingdom*, while cells without chlorophyll belonged to the *animal kingdom*.

(This is the simplest way of defining the difference between the two kingdoms, and its very simplicity makes it, almost inevitably, misleading. Bacteria, yeasts, molds and various fungi, which do not contain chlorophyll, are nevertheless generally included in the plant kingdom. Some classifiers try to increase the logic of the classification by lumping the simpler microorganisms into a third kingdom called *Protista*. Some even divide one-celled creatures into two separate kingdoms, making four altogether. I shall stick to the two-kingdom system, however, and speak of plants and animals only.)

The development of plant cells meant a vast change of life for those one-celled organisms that did not develop chlorophyll and which we can now refer to as animal cells. The animal cells no longer depended on the scum of natural foodstuffs in the ocean. Now that plant cells were creating new quantities of food out of air, water and sunlight, animal cells that learned to engulf plant cells and strip them of their stored food entered a new (and far richer) environmental niche. These animal cells replaced the more old-fashioned type, and the general pattern was established that still persists today. Plants form and store food, while animals eat plants (or other animals that have eaten plants, or other animals that have eaten still other animals that have eaten plants, and so on).

It might seem that of the two kingdoms, the plants chose the better and more independent path. After all, plant life in some of its forms can continue to exist indefinitely, even though all animal life were destroyed, but the reverse is not true. No animal life today would exist for more than a comparatively short period after the destruction of all plant life (unless the primordial conditions whence arose a food supply in the ocean were to be reproduced, which is extremely unlikely). In fact, animals may almost be thought of as parasites upon the plant kingdom.

And yet this seeming independence of the plant is an illusion. It is the plant, in fact, that bears the mark of parasitism. The true parasite is one that lives in its food supply; its food is also its environmental niche. Many animals do indeed qualify as parasites under this definition; but most do not. Animals must seek their food, sometimes fight it (or for it) and lose out.

It is the plant kingdom, generally, that lives within its food supply. It is surrounded by the air and sunlight, two of the essential factors it needs, and ocean plants are surrounded by water, the third factor. Surrounded by its food, the plant kingdom need not struggle in the same sense that animals do (what struggling goes on is largely passive).

Animals have generally had to develop many specializations in order to compete among themselves for a food supply infinitely harder to obtain than is the case among plants. To cite just one example, animals generally had to develop the trick of independent motion, which plants in general have never had to do.

The possession of mobility (to say nothing of other specialized traits) enables animals generally to control their immediate environment more efficiently than plants, as a rule, can. Animals more actively engage in forcing their environment to accede to their demands or to keep it from harming them. To put it as simply as possible: a rabbit can bite a carrot, but a carrot cannot bite a rabbit.

If we wish, we can define an organism as being more "advanced" when it can more effectively control its immediate environment. By this definition, animals generally are more advanced than plants.

Continued competition among cells would lead naturally to the development of more and more effective control of the immediate environment on the part of the varieties that survived natural selection. That greater control of the environment would, in fact, be the reason for their survival.

The greater control would almost inevitably involve an increase in size. The larger a cell, the greater the variety of chemical substances it can contain and the more chemical tasks it can perform. The larger a cell, the more food it can store, the more energy it can generate, the more it can partition itself off into specialized subdivisions. In short, a large cell can do more than a small cell can, and is bound to be, by our definition involving increased environmental control, more advanced.

But as cells grow larger, troubles arise. The rate at which food enters a cell and wastes depart depends on the surface area of the cell. The total food requirements of a cell depend, however, on its volume. As a cell increases in size, the volume increases as the cube of the diameter, the surface only as the square. If a cell's diameter is doubled (assuming a spherical shape), its area, through which it must feed, is increased fourfold, while its volume, which does the feeding, is increased eightfold. If the spherical shape is maintained, a size is quickly reached in which there is no longer enough surface to feed the increased bulk.

An alternative would be to abandon the spherical shape. Cells might grow long or flat or irregular, all in order to increase the surface without increasing the volume. Some undoubtedly tried this; there are primitive life forms existing today, called slime molds, which are single, flat masses of protoplasm that can be as much as ten inches across. This is not a very successful variety of life, but slime molds survive, just the same.

A second alternative is for cells to remain small and reasonably spherical, but to stick together after cell division. In this way, groups of cells are formed which, by hanging together, have whatever advantage sheer mass brings them. For instance, they would be less at the mercy of wave and current, and by united action could swim more strongly. While these advantages accrued, each individual cell would remain within the safety limit of the "square-cube" law.

Such groups of cells are called *cell colonies,* and they can be made up of plant or animal cells.

An example of a cell colony, formed of numerous plant cells, is Volvox, in which the cells form a hollow sphere about a fiftieth of an inch in diameter. Each cell has *cilia* (tiny, hairlike projections, from a Latin word meaning "eyelash") which beat in coordinated fashion and allow the colony to move through water by rolling. ("Volvox" comes from a Latin word meaning "to roll.") Each cell in a Volvox, however, is just about as it would be if it were living independently.

When a cell forms part of a cell colony, its environmental niche changes fundamentally. The other cells of the colony become a major portion of the niche. The individual cell can fit itself to the surrounding cells best by exaggerating some one of its own functions for the benefit of its neighbors, even though this might be to its own detriment as an "all-round" organism. Its deficiencies would be made up for by the compensating specializations of the cells surrounding it.

One analogy that may make this plain is the comparison of a group of workers in a factory with an artisan working at home. The artisan, manufacturing an article, sees it through every step and must be able to perform each one with adequate, if not superlative, skill. The worker in a factory—who may be considered as part of an "artisan colony"—can afford to specialize in a single step of the process, learning to do that particularly well, even though his ability to do the other steps deteriorates. Other workers will have their own specialties, all of which will complement one another and fit into a unified whole. The end result, as proven by the economic history of the last two centuries, is the greater efficiency of the factory (or cell colony) over the home workshop (or individual cell).

Among the most specialized, and therefore most successful, cell colonies are the sponges, a collection of animal cells which can grow to quite large size. Sponges are made up of several types of specialized cells, each of which performs its particular job for the benefit of the whole.

There is a type of cell that secretes a gelatinous, fibrous material that both supports and protects the sponge. The colony, through a combination of this and its great mass, is safer and better protected against the stresses of the environment than any individual cell can be. Since improved protection against the environment is one way of successfully coping with it, a sponge is, by that fact alone, at once seen to be more advanced than a single cell can be.

Other sponge cells have *flagella* (from the Latin word for "whips") which are whiplike threads, longer than cilia. These can stir up a current by their lashing, a current that will carry food particles into the colony and expel wastes. Still other cells contain pores through which the current will pass. These pores are so characteristic that the sponges are placed in a separate phylum of their own, called *porifera,* from Latin words meaning "pore-bearers."

Yet in a cell colony even as complicated as the sponges, the individual cells have not entirely given up their birthrights. Their increased specialization does not prevent them, at need, from fulfilling the other functions of a cell. Each sponge cell is capable of independent life and any one of them may, and sometimes does, wander off on its own to start a new colony.

But suppose we imagine this trend of specialization carried on to a logical conclusion. To increase the efficiency of a cell colony, more and ever more specialization is required. Each cell must get better and better at its particular task even if it means that other abilities are allowed to become completely vestigial.

Eventually the individual cell of a colony becomes so

specialized that it can no longer exist on its own; it can only stay alive as part of a group. It has, in a sense, become parasitic on the colony and unfitted for any environmental niche other than the colony. When this point is reached, we are dealing with more than a cell colony: we are dealing with a *multicellular organism*.

In the less advanced multicellular organisms, specialization has still not gone all the way. Though individual cells are helpless on their own, relatively small groups of cells can, if torn loose, survive and serve as the nucleus of a new organism. This is *regeneration*. As multicellular organisms advance through ever increasing specialization, however, powers of regeneration grow progressively less.

A multicellular organism can, by the fact of the greater specialization of its component cells, control its immediate environment more efficiently and more extensively than a cell, or even a cell colony, can. A multicellular organism is, therefore, more advanced than any single cell or cell colony can be. This increase in advancement, however, exacts its price. For instance, a multicellular organism pays for its advance by loss of immortality.

An individual cell is potentially immortal. Given sufficient food and safety, it will grow and divide forever. The multicellular organism, however, depends not only on the cells that compose it, but on the organization among them. The malfunction of a few cells may destroy that organization or a vital portion of it, and bring death to the entire organism, including all the healthy cells that compose it. (To put this in simplest terms, a blood clot in the brain may be fatal, though no other part of the body has been directly interfered with.) And, of course, it is common experience that, sooner or later, intercellular organization will break down and, despite ample food and care, all multicellular organisms, even men and redwood trees, are mortal.

It is interesting to note that plant cells, with their easy life and their parasitism on sun, air, and water, made the advance into multicellularity much later than did animal cells; and when they did, the advance was much less extensive and intensive than in the case of animals. In fact, the plant life of the sea never advanced beyond the cell-colony stage at all. The most elaborate seaweed is only a cell colony.

It was only when plants invaded dry land, and when water and minerals consequently became less easy to obtain, that groups of plant cells had to be specialized to the point where true multicellularity existed. Groups of plant cells were developed, for instance, for the prime purpose of absorbing water and minerals out of the ground, others to collect light from the sun, others to support the plant, and still others to communicate water from below and food from above to other parts of the organism. Even so, the most elaborate tree is not as elaborate as even a simple animal. No plant, for instance, has a nervous system, muscles, or a circulating blood system. And even advanced plants retain greater powers of regeneration than do all but the simplest animals.

We can restrict ourselves to multicellular animals now, and, what is more, feel no need to concern ourselves with each of the more than twenty phyla into which taxonomists usually divide them. Although species of each phyla survive to this day, those of at least half occupy out-of-the-way environmental niches. To present the general scheme of advance, we need consider only the nine phyla that represent the main landmarks of evolutionary development.

To begin with, the least advanced of the multicellular phyla is that of the *Coelenterata*. Familiar living examples of this phylum are the fresh-water hydra and the jellyfish.

The coelenterate body plan, in simplest terms, is that of a cup shaped out of a double layer of cells, both layers containing specialized members. The layer facing the outside world is the *ectoderm* (Greek for "outside skin"); that on the inside of the cup is the *endoderm* ("inside skin").

The ectoderm, dealing primarily with the outer world it faces, contains primitive nerve cells to receive and transmit stimuli, thus coordinating the behavior of the component cells of the organism. It also contains stinging cells that serve as weapons of offense.

The endoderm, on the other hand, is a food-centered layer. It contains cells specialized to secrete juices into the cavity of the cup, where food particles are digested and prepared for absorption. The key advance made by the coelenterates is the possession of the interior of the cup as a private bit of the ocean. In cells and cell colonies, however complicated, food particles must be engulfed by individual cells before they can be made use of.

The coelenterate can, instead, pop food particles into the interior of the cup (into the "gut," that is) after having seized those particles with the tentacles that rim the mouth of the gut. After digestion has taken place within the gut, the cells of the endoderm need only absorb the dissolved products of digestion, not the particle itself. In this way, many food particles can be handled at once, and individual food particles considerably larger than an individual cell can be taken care of. The very name of the phylum emphasizes the importance of this feature of its being, since "Coelenterata" comes from Greek words meaning "hollow gut."

The next major step in development was the formation of a third layer of cells, a *mesoderm* ("middle skin") between the ectoderm and endoderm. Out of the mesoderm, whole systems of specialized organs (groups of cells designed to perform particular jobs) can be developed. Each system of specialized organs increases the efficiency with which, in one way or another, the organism can control its environment, so the presence of the third cell layer is an enormous advance.

This third cell layer seems, in all probability, to have been developed on two different occasions from two different primitive groups of coelenterates. At least, in one group of the higher phyla, the mesoderm arises from the junction point of ectoderm and endoderm, while in a second group, the mesoderm arises as an outcropping at various places in the endoderm. There are other general differences between the two groups of phyla as well. In consequence, some classifiers group all organisms more advanced than the coelenterates into two "superphyla," the *Annelid superphylum* and the *Echinoderm superphylum*. First, let us consider the Annelid superphylum, which is the larger (though, as it happens, it is not the one to which we belong).

The most primitive phylum of the Annelid superphylum is *Platyhelminthes* (Greek for "flat worms," actually referred to very commonly as "flatworms" in English).

The flatworms use the mesoderm (which, in the history of life forms, they were probably the first to possess) to form contractile fibers, which are the first muscles. Out of the mesoderm, the flatworms also formed special reproductive organs and the beginnings of excretory organs.

In addition, the flatworms were the first creatures to display *bilateral symmetry*. In single-celled creatures, in cell colonies and among the coelenterates, what symmetry there was, was radial. That is, if a line is drawn vertically through the center of a jellyfish, for instance, it would form a kind of axis about which the jellyfish is the same in all directions, like a wheel.

In creatures with bilateral symmetry (including almost all creatures from flatworms to man) a line down the center of the body from front to back divides the animal into a left and right side that are mirror images of each other, while the front and rear ends are not mirror images. It is this, in fact, which gives a creature a "head" and a "tail."

The advantage of bilateral symmetry is that motion, whatever the direction, is generally "head-first." The fact that it is the head that specializes in poking into the unknown means that a number of sense organs can profitably be concentrated in that area. Concentration of sense organs into one spot leads to greater efficiency of sense perception generally, to a consequent better control of the environment and, by our definition, is an evolutionary advance.

For example, flatworms have further developed the primitive nerve cells of the coelenterates and include a centralized nerve "cable," with a concentration in the head area, where it is most needed. This is the beginning of the first nerve cord plus brain.

Both coelenterates and flatworms must remain small or, at most, can grow in only two dimensions. Food, when absorbed from the primitive gut, must be passed along from cell to cell. No cell can afford to have too many layers of cells between itself and the gut, or not enough food will reach it (too many middlemen, too many greedy mouths between). The flatworms improve over the coelenterates by possessing, in some cases, a branched gut about which more cells can huddle. This is only a palliative.

To be sure, the coelenterates include giant jellyfish, but their long stingers are very thin and their voluminous "bell" is composed mainly of a very watery, jelly-like material (hence "jellyfish"), with the actual living cells

concentrated near the surface. There are also giant flatworms, such as seventy-foot-long tapeworms, but these are
flat as tape, as the name implies, and not wide, so that all
the cells are near the surface (which is absorbing food
from the intestine of the creature it is parasitizing).

To enable a multicellular organism to achieve real bulk,
as distinguished from simple length, some method must
be devised for distributing food to, and withdrawing wastes
from, cells which may be deeply buried under numbers
of cell layers so that they are far from either the outside
surface or the gut surface of the creature.

Such a development is to be found in the phylum
Nematoda (from Greek words meaning "thread worm,"
popularly called "roundworms" in English). The nema-
tode development is that of a fluid within the mesoderm
layer, which can slosh back and forth through the nooks
and crannies of the organism, bathing all cells. Food and
oxygen can now be poured into this fluid by those cells
which absorb an excess from the gut, or from the outside
world. The fluid (a primitive blood stream) can distribute
it evenly to all cells. It can also act as a repository for all
cell wastes, which it can carry to the excretory system.

In short, the nematodes developed an internal bit of
ocean, so to speak. While a cell has an "ocean front" on
the blood stream, as all have, it need not worry about a
real ocean front. That is why the nematodes can develop
bulk (though most, to be sure, remain quite small) and
be round worms, while the members of *Platyhelminthes*
could not, and had to remain flat worms.

The nematodes are also responsible for another advance. In both coelenterates and flatworms, the gut has
only one opening. The indigestible residue of food that
has been taken in has to be ejected through the same
opening by which it originally entered. While ejection is
taking place, further ingestion cannot, and vice versa.

The nematodes added a second opening to the gut, one
in the rear (the first *anus*). Food particles enter at one
end, are digested and absorbed as they travel along the
gut, and the indigestible residue is ejected at the other
end. Both ingestion and ejection can be continuous, and
obviously this moving-belt, assembly-line technique of
feeding (which has been retained in all higher phyla)
represents another major improvement in the control of

the environment and, consequently, another evolutionary advance.

Many phyla of animals have at one time or another developed hard or tough frameworks that serve to protect them against the force of waves and currents, as well as against the onslaughts of enemies. A number of one-celled organisms have developed limy or siliceous shells; cell colonies such as the sponges have also developed a tough skeletal framework. A group of coelenterates, popularly known as corals, are small creatures living within incrusting lime skeletons in huge colonies that are reminiscent of sponges.

The first creatures, however, to convert these hard incrustations into definite shells which could be maneuvered to suit the needs of the animal were the members of the phylum *Brachiopoda*. These have two equal shells, made of limestone, one on the top and one on the bottom. These shells can be brought together or moved apart by means of muscular attachments; they resemble ancient lamps, so that the creatures are popularly called "lampshells." The brachiopods are equipped with stalks by which they are attached to rocks, and these may be viewed as an arm by which they hold on, or a leg on which they stand. In any case, the word "Brachiopoda" comes from Greek words meaning "arm-leg."

The internal organs of the brachiopods are more highly developed than those of any of the various worm phyla. A phylum with internal organs still more developed is that of the *Mollusca*. These also have shells of limestone, which, however, are unequal in size and originate from the right and left sides of the creatures, instead of top and bottom. (Such differences are of the type that make taxonomists feel warranted in placing the mollusks and brachiopods in two different phyla. The word *"Mollusca"* comes from a Latin word meaning "soft," since within the hard external shell there is a soft body.)

The development of shells not only increases safety, but also makes possible more efficient muscles, since the shells offer a firm attachment point for muscles. The shells also allow for considerable bulk. There are mollusks such as giant clams, for instance, that are six feet across.

However, the development of shells involves serious shortcomings, too. The dead weight of the armor deprives creatures wearing it of the mobility so painfully developed by animals generally. The shell also serves to block off the creature from the outside world, reducing the variety of its sense impressions. In place of the wriggling, active worms, you have the motionless oyster. The mollusk and brachiopod achieve security by means of an eternal passivity; whether the net result is considered an advance is a matter of taste, perhaps.

Indeed, the most successful of the mollusks are grouped in the class *Cephalopoda* (from Greek words meaning "head-feet," because the tentacles or "feet" of creatures in this class seem to spring directly from their heads). The cephalopods, including such creatures as squids and octopi, are successful precisely because they have largely abandoned the shell, while retaining the particular specializations that go with the new bulk first made possible by the shell. The giant squid in particular (which may be thirty or forty feet long, if its tentacles are included), with its bulk, its large eyes (a mollusk development), its powerful tentacles, and its method of moving by jet propulsion, is perhaps the most specialized and successful form of simple, "one-piece" life that ever existed.

What is meant by "one-piece"? We'll come to that next.

Branching off from the nematodes in a direction different from that taken by the brachiopods and mollusks is still another "worm" phylum, the *Annelida*. The annelids never developed shells and are merely wormlike in appearance, but they did make an advance that was perhaps the most important after the establishment of multicellularity. (Because of it, they give their name to the Annelid superphylum to which they belong.)

The advance is that of *segmentation*. An annelid is composed of a series of segments, each of which may be looked upon as an incomplete organism in itself. Each segment has its own nerves branching off the main nerve stem, its own blood vessels, its own tubules for carrying off waste, its own muscles, and so on. Such a body plan is again an example of the assembly-line philosophy that had originally led to multicellularity. It is, in fact, a kind of half-step beyond multicellularity.

The giant squid, the most advanced of the "one-piece" animals, is the most advanced form of nonsegmented life that ever existed. Although the annelids are not to be compared with the giant squid in bulk or efficiency, they have, by virtue of their segmentation, a potentially greater flexibility and specialization than any of the lower phyla up to and including the mollusks.

The best known of the annelids is the common earthworm. Anyone looking at an earthworm can see that there is a series of constrictions encircling its body down its entire length. These constrictions are the boundaries between the segments; they make the earthworm look as though its body were composed of little rings of tissue fused together. (The word *"Annelida"* is from a Latin word meaning "little rings.")

Perhaps because of the efficiency implied in segmentation, the annelids, although the least advanced of the segmented phyla, were able to introduce new specializations. For instance, they improved the circulatory system by enclosing it in tubes (blood vessels) and driving the current by means of a pump (the heart). They also developed hemoglobin, a protein which could carry oxygen with far greater efficiency than could a simple watery fluid.

Yet the annelids lacked a skeleton. They remained soft, relatively defenseless, and limited in potential bulk. (Even the famous six-feet earthworms of Australia remain long, thin and unbulky.)

The obvious next advance, then, is to combine the efficiency of segmentation with the security made possible by a skeleton. From the annelids there descended those creatures belonging to the phylum *Arthropoda*. The arthropods, which include such creatures as lobsters, spiders, centipedes and insects, possess an outer skeleton.

The arthropod shell is, in many ways, a vast improvement over the shells of the brachiopods and mollusks. The arthropod shell is not limestone but a light, tough, flexible substance called *chitin*.

Moreover, the arthropod shell is more than a single-piece barrier against the outside world. It is segmented, as is the rest of the arthropod body and, therefore, fits the contours of the body and limbs closely. It is the jointed limbs that give the phylum its name, for *"Arthropoda"* comes from Greek words meaning "jointed feet."

The arthropods are the peak of evolution within the Annelid superphylum. Its most successful class, *Insecta*, dominates the world, so far as the number of species and of individuals are concerned. There are more species of insects than there are of all other animals put together. The weight of insects on the earth is greater than the weight of all other animals combined (including whales and elephants). Insects flock into every environmental niche, compete for food all too successfully, even with man, at every stage; they will very likely survive man.

But let us return to the coelenterates and consider the second group of phyla, the Echinoderm superphylum, which developed from coelenterates through the formation of a mesoderm from outgrowths of the endoderm.

The first phylum so developed was *Echinodermata* (Greek for "spiny skin"). This includes such creatures as the starfish and the sea urchin. Like the flatworms, which had developed a mesoderm independently, the echinoderms developed bilateral symmetry, but, unlike the flatworms, they abandoned it in favor of a return to radial symmetry. Bilateral symmetry can still be seen in the infant forms (or larvae) of the echinoderms.

The echinoderms also developed a kind of shell that was fundamentally different from those developed by brachiopods, mollusks and arthropods. The echinoderm shell was formed under the skin so that it was a kind of internal skeleton.

Nevertheless, the echinoderms themselves did not go as far as did either mollusks or arthropods. The radial symmetry kept their sensory equipment retarded, while their shells weighed them down and kept them relatively passive.

But from the echinoderms there developed a second phylum which was much more important. It arose early, after the primitive echinoderms had developed bilateral symmetry and before they abandoned it again. The new phylum kept bilateral symmetry. The new phylum also lightened the internal skeleton, freeing itself of the dead weight. They kept only one piece of it to begin with, a single, stiffening rod of gristle along the back, called a *notochord* (from Greek words meaning "back string"). This served to protect the main nerve fiber which ran along the back parallel to the notochord (and not along the belly, as in other phyla).

More important still, the new phylum independently

developed segmentation—not one as thoroughgoing as that developed by the annelids, but segmentation, nevertheless. Thus, the new phylum kept what was good, abandoned what was bad, and developed what was better. It was headed for great successes.

This new phylum (the latest to be developed in the history of life) is the *Chordata* (from *"notochord"*) and it is the one to which *Homo sapiens* belongs. Therefore a diagrammatic summary will be included of the development of the phyla, as outlined in this chapter, and then we will pass to a new chapter dealing with the *Chordata* in some detail.

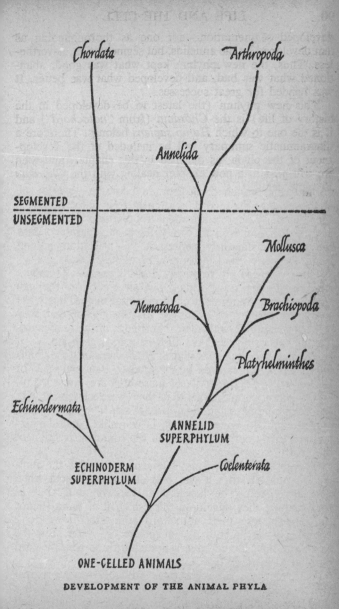

DEVELOPMENT OF THE ANIMAL PHYLA

7
Inside Chordata

The chordates evolving from very early forms of the echinoderms (perhaps both chordates and echinoderms branched off from a common ancestor of which we have no record) did so in four different directions. These directions are today represented by creatures so different that even to group them as separate classes does not satisfactorily represent the variation among them. They are listed in four different *subphyla*.

Three of these are relatively undistinguished. Members of each of them exist today, but all are out-of-the-way creatures inhabiting quiet environmental niches. They are in the backwaters of evolution for the reason that they have done nothing with their inheritance of the notochord. Two of them, in fact, have allowed it to degenerate.

The first subphylum is that of *Hemichordata* (the "half-chordates"), represented by the modern Balanoglossus, or "acorn worm." The larvae of these creatures are so similar to echinoderm larvae that they were classified with the echinoderms when first discovered. The adult form, as the common name implies, is wormlike, and the notochord exists only as a scrap in the foremost fraction of the creature.

The second subphylum, *Tunicata*, resembles the mollusks so closely that it was classified among them when first discovered. The larvae are free-swimming, but as they mature, they develop a thickish shell or tunic (hence the name), lose their notochord and settle down to a completely sessile life, like a sponge or oyster. But the free-swimming larvae have a notochord and a nerve cord parallel to it.

Both of these subphyla were probably more progressive to begin with, but found the environmental eddies they were inhabiting taken over by still more advanced forms; so only those species survived which had retrogressed to make use of wormlike or mollusk-like niches.

The third subphylum, the *Cephalochordata,* includes creatures that have retained notochords throughout adult life. In fact, the notochord runs the full length of the body into the head, whence the name of the subphylum, which in Greek means "head string." The best known of this group is the amphioxus, a creature some two inches long that is fishlike in appearance. It clearly shows its musculature divided into segments.

It may seem odd to think of the chordates (including human beings) as segmented, since a casual glance at familiar members of the phylum does not show it. However, in eating fish, you will notice that the meat comes away easily at certain dividing points, which are the boundaries between segments. In the human body, anatomists can easily note the manner in which nerves, muscles and blood vessels repeat themselves in each segment.

The segmentation, to the untutored eye, shows up most clearly, however, in the skeleton. Feel your backbone and you will discover a series of individual bones, called *vertebrae,* running from just under the skull to the base of the spine. Each vertebra represents one segment. This is most dramatic in the chest region where each segment possesses a pair of ribs as well. (Or look at the skeleton of a large snake, where vertebrae and ribs have been multiplied, and see if that example of chordate skeletal construction does not remind you of the clearly segmented centipede.)

If the three subphyla so far listed were the only ones to represent the *Chordata,* the chordates would certainly have to be marked down as among the least successful of the phyla. However, there is a fourth subphylum, *Vertebrata,* so called because its members developed the notochord into the vertebrae just mentioned. The vertebrae made it possible for the back-stiffening to be more flexible and stronger than the simple, rodlike notochord. Furthermore, along with the vertebrae there came the key development that made the *Chordata* an inevitable success. Stiffening rods of cartilage formed the ribs, supplying the body with an internal framework, from which muscles could be suspended and behind which soft organs might be protected.

The internal framework was at once lighter, more flexible, and more versatile than the outer shells of other phyla, which achieved results by sheer mass and surrendered to deadening weight. The importance of this internal framework is such that earlier zoologists divided all the animal kingdom into two parts: the *vertebrates* and the *invertebrates*. Among specialists, this has been replaced by the division into phyla, but in popular speech this two-way classification still holds.

The subphylum *Vertebrata* is divided into eight classes which are in turn collected in two groups, each group comprising a *superclass*. The first superclass is *Pisces* (the Latin word for "fishes"), which includes those vertebrates whose home is primarily in the water. The four classes within *Pisces* show a progressive improvement in the skeleton.

The first class is *Agnatha* (from a Greek word meaning "jawless"). The agnaths were the first to develop vertebrae and ribs, the essential portions of the internal framework, but they lacked jawbones. The best-known modern representative is the lamprey or hagfish which, in place of jaws, possesses a rasping set of files in a round, sucker-like mouth. It can attach itself to a fish, rasp away the skin and devour the interior. (The lamprey is the nearest thing to a parasite among the chordates.)

The agnaths were not only the first to develop the skeleton; they also developed eyes (which mollusks had done independently) and improved the circulation by the development of a two-chambered heart and of red blood cells in which to keep hemoglobin (which annelids had done independently).

The next class, *Placodermi* ("plate skins," because they were covered with platelike, bony armor), converted one of the gill arches into a jaw and were the first vertebrates with a mouth that could open and close. This class is entirely extinct now, having been replaced by the more efficient classes of *Pisces* that overlapped the environmental niche of the placoderms too thoroughly.

The *Chondrichthyes* (represented today by the sharks) added four limbs, each equipped with internal skeletal bracing, along with teeth in the jaws. The limbs, which were stubby and ended in fins that served as paddles (plus balancing fins and a fin-tipped tail), made for faster and more efficient locomotion, while the teeth made for more effective offense and feeding.

The skeleton was now essentially complete, and the success of the *Chordata* became quite marked. The sharks are bulkier and more efficient by far than any of the nonchordates of the sea. The largest shark (the whale shark) may be up to sixty feet long; and although giant squids may approach this length, most of that length is made up of long, thin tentacles, whereas the shark's body is thick and massive.

The fact is that the bracing of an internal skeleton (an *endoskeleton*) is more efficient as a strengthening agent than is the solid embrace of an external skeleton (or *exoskeleton*). The exoskeleton is a passive shield, a knight's armor. The endoskeleton is a well-designed girder system, similar to the steel beams of a skyscraper.

An exoskeleton limits growth. If the soft inner tissues grow, they are restricted by the slow and clumsy growth of the outer shell. In the arthropods, the barrier is short-circuited by the periodic discarding of the shell and its replacement by a new and larger one. A great deal of vital energy goes into the perpetual manufacture of such exoskeletons. Furthermore, during the interval between the discarding of the old skeleton and the hardening of the new, the arthropod is comparatively defenseless.

An endoskeleton, however, does not limit growth. The bones within may freely be extended by accretion and the soft tissue about it yields and matches the growth easily.

There was, however, one improvement still to be made in the skeleton; not so much in its extension as in its constitution. The first three classes of *Pisces* had skeletons made up of *cartilage* (the flexible gristle that stiffens your outer ears and the tip of your nose). The fourth class, *Osteichthyes,* also developed from the placoderms but did not entirely discard the notion of bony armor. The armor they abandoned, but the bone they moved inside. Their skeleton was converted from flexible cartilage into hard bone. (In fact "*Chondrichthyes*" is Greek for "cartilaginous fish" and "*Osteichthyes*" for "bony fish.")

A bony skeleton is stronger than a cartilage skeleton, but the advantage is not so great that bony fish entirely replaced sharks. Both classes flourish today and will undoubtedly flourish indefinitely in the future.

The greatest thing about a bony skeleton, however, is that for the first time an internal framework had been evolved that was strong enough to enable bulky animals to invade the land.

The conquest of land was forced upon life by the pressure of circumstances. Life undoubtedly developed in the surface layers of the ocean, and most of it is still there. Something like five-sixths of the total mass of living matter dwells in the surface layers of the ocean even today. Creatures who live elsewhere did not go voluntarily; they were pushed by the necessity of finding new environmental niches in the face of competition by more efficient organisms taking over occupation of their old ones. Those that did not manage (like the placoderms) became extinct. Those that managed remained alive, and sometimes went on to new successes.

Some creatures, for instance, were forced to invade fresh water, a much less desirable niche than the ocean because the fresh water of rivers and lakes is poor in necessary minerals. To live in fresh water, creatures had to develop chemical mechanisms for collecting and conserving the rare minerals. All chordates possess these mechanisms, and from this it is deduced that the original chordate must have developed from a primitive echinoderm (or pre-echinoderm) ancestor in fresh water, although some of the chordate descendants migrated triumphantly back to the sea.

Other creatures of the ocean surface were forced into the less and less desirable niches of deeper and deeper water, where plants do not exist and food is harder to find. Still others were forced into stagnant, marshy water.

But land was certainly the least desirable niche of all. In the absence of water buoyancy, gravity makes itself felt in full force, so that there is a constant and terrible pull downward. There are temperature extremes: scorching heat in the full blast of sunlight and freezing cold in the winter night, in contrast to the equable and even temperature of the ocean. There is the lack of water which imposes the constant danger of death by drought, and there is the frightening force of the wind.

Such, however, is the pressure of competition that creatures were forced into tidal waters, where they were periodically exposed to land conditions, then further and further up the beach until finally they were land organisms altogether.

The first life forms to manage the transition were, as was to be expected, plants. This took place about three hundred and fifty million years ago, not long after the first vertebrates evolved in the oceans. These pioneering plants, belonging to a group called *psilopsids,* were the first multicellular plants, the plant equivalent of the coelenterates. They did not develop roots, but they developed stems, and some even developed simple leaves. (Almost all the psilopsids are now extinct and have been replaced on land by more complex plant phyla.)

Once plant life had established itself on dry land, there was an opportunity for animal life to follow suit, for by then, you see, the land offered animals a food supply. Within a few million years, small arthropods followed: scorpions, spiders, primitive insects. There were also snails and worms, all feeding on the psilopsids and on each other.

All these creatures were alike in being small. Creatures without skeletons, or with exoskeletons only, could not fight the gravity of a waterless environment unless they were small. In fact, an exoskeleton is a disadvantage, since the weight it adds is no compensation for the bracing it offers. The slowness with which a land snail moves (its shell on its back) is proverbial.

The first really mobile land creatures were the insects. They managed this by learning to swim through air (fly) instead of water. To do this, they had to develop filmy wings, much like tiny fins in appearance. Because air is so much less buoyant than water, wings had to be moved much more rapidly than fins.

So the situation remained constant for the first hundred million years after the original invasion of land by life. Then, just about the time the insects conquered the air, something new came to the fore—the internal bracing of bone.

The bony fish that then moved out on land were an unsuccessful variety that found the competition in the ocean's surface too difficult. These fish belonged to the subclass *Crossopterygii* (Greek for "fringed fins"). They were marked by limbs that were scaled and only fringed with fins. Either because these were less efficient for swimming or for other reasons, the crossopterygians could not stand up against the other subclasses of bony fish.

Most of the crossopterygians are now extinct. A couple of species, with fins that have grown almost vestigial, have developed into the lungfish of Africa and Australia. These exist in stagnant water, in which ordinary fish would suffocate. The lungfish even survive through summer droughts when the water dries up completely. This is made possible by the fact that most fish possess a swim bladder and that the lungfish were able to convert that bladder into a lung. (The swim bladder is an air-filled vessel which can increase or decrease the buoyancy of a fish by the change in the volume of air it contains. The fish can use this device for vertical travel, much as a submarine makes use of precisely the same device.) Whereas most fish absorb dissolved oxygen from water by means of gills, the lungfish can gulp gaseous air into its swim bladder, dissolving some of the oxygen into the moisture that lines the bladder, and survive.

Other crossopterygians were forced into the abyss, and a few species are still alive there. One specimen was brought up in 1939, to the dumfoundment of all zoologists who considered the entire subclass to have become extinct seventy-five million years ago.

Finally, about two hundred and fifty million years ago, some of the crossopterygian fishes, moving painfully and sluggishly on their four sturdy fins, invaded the land. The internal framework of bone could prop up even large masses against the pull of gravity and hold it firm against the push of the wind.

Often environmental niches that are entered with reluctance are really undesirable under any conditions, and those who are trapped therein can do very little more than survive. The oceanic abyss is an example of this—a complete dead end. Dry land would seem to be even more of a dead end, but instead it turned out to be the environmental niche of the future.

The chief reason for this is that air is about seventy

times less viscous than is water, and offers that much less resistance to motion.

A creature capable of rapid motion is, generally speaking, in better control of its environment and, therefore, more advanced (all other factors being reasonably equal) than one not capable of rapid motion. This holds true on sea as well as on land, and the most advanced sea creatures are indeed capable of rapid motion.

But for a sea creature to move quickly, streamlining must exist; otherwise an impractical amount of energy is consumed in overcoming water resistance. The streamlining of sharks and bony fish is an obvious example.

Creatures on land, however, may be designed for rapid movement through the much less viscous air without being quite as extensively streamlined.

The difference involved is enormous, for streamlining makes the effective use of appendages impossible. Appendages of any length would break up streamlining and destroy efficiency of motion. Yet it is precisely by means of appendages that creatures can best handle environment and bend it to their will—by prehensile tails or maneuverable trunks, by powerful running limbs or delicate hands.

Consider the whale. Large whales have brains that are larger than those in a man's skull, and more convoluted. Can the whale be as intelligent as a man? Supposing it were, how could we tell? It has a tail and two flippers with which it can do nothing but swim. It has no appendages with which to manipulate the outside world and, because of the necessity for streamlining, can have none. What intelligence the whale might have must remain potential—a prisoner of the viscosity of water.

The giant squid has tentacles that drag out behind when it is in rapid motion and which do not interfere with streamlining, yet which remain useful appendages when it is not in rapid motion. However, those tentacles can move only in slow motion, thanks to the viscosity of water. (Try swinging a bat under water and you will see what is meant.)

To summarize, then, the appendage is rare in the sea, and the quickly moving appendage is nonexistent. The quickly moving appendage is, however, common among land creatures, and it is that which makes land species, in general, more advanced by our definition than sea species.

Four classes of Vertebrata inhabit the dry land, forming a second superclass, *Tetrapoda* (from Greek words meaning "four-footed," the Latin version, "quadruped," being more familiar).

The first class of Tetrapoda (which would include the first crossopterygians who made the transition) is that of the *Amphibia* (Greek for "double life"). The frogs and toads are the best-known modern representatives of this class. Amphibian lungs, working full time in adult life, needed a circulatory branch of their own. A three-chambered heart became a necessity. In addition, the amphibians developed the ear. Both are evolutionary advances over anything in Pisces.

But amphibians were (and still are) tied to a watery environment of some sort during the period of birth and early development. Eggs must be laid in water and the young amphibia breathe by gills, so that each individual must repeat the victory of the class and conquer the land for itself (whence the "double life" reference in the name of the class).

It was the next class, *Reptilia* (the modern representatives of which are snakes, lizards, alligators, turtles and so on), that made the conquest of land complete by developing a crucial improvement—an egg that could be hatched on land. This was accomplished about two hundred and twenty-five million years ago; or some twenty-five million years after the first amphibian crawled out onto land.

Such a land-based egg had to be enclosed by a shell which was porous to gases (so that the developing embryo could breathe) but which would retain water so that the embryo would not die through drought. The egg had to be large enough to contain the food and water needed by the embryo through a period of development long enough to enable it to reach the point of independent life on dry land. This meant the embryo had to develop special physical and chemical devices whereby it might handle the food contents of the egg.

The reptiles achieved such eggs and became true land animals at all stages of their lives. They also put the final touches on the circulatory system by developing a fourth (and last) chamber of the heart so that two complete and coordinating blood-pumps existed. Furthermore, the reptiles developed stronger legs that could lift the body clear of the ground. They could move more surely and more quickly, and that alone gave them an advantage over the amphibians, most of whom they replaced. The reptiles reached their heights from two hundred million to one hundred million years ago, during which time the giant dinosaurs developed and for a time shook the earth.

(Oddly enough, despite their development of strong legs, modern reptiles have regressed in that respect to a large degree. In fact, the most recently developed and most successful of the modern reptiles are the snakes, which have abandoned legs altogether. The very words "reptile" and "serpent" come from Latin words meaning "to creep.")

Amphibians and reptiles conquered the gravity of dry land and the lack of water, but there still remained the problem of temperature variations. Reptiles developed many devices to insure an equable internal temperature. They sunned themselves when the sun was out (and still do in modern times). Some large reptiles, now extinct, even developed huge sails along their back, held up by spines, which probably acted to catch the sun. This served to pick up heat that might be carried by the blood stream to the rest of the body. One modern lizard sticks the top of its head out of its burrow, allowing a large blood vessel just under the skin to soak up the sun's rays and carry warmth to the rest of the body. Only when its body temperature rises high enough does it come out. Then, of course, when the temperature is too high or too low, reptiles can always remain in caves, burrows or crevices.

Yet the best devices remained comparatively makeshift, particularly for large reptiles which could not as easily get under cover. In fact, there is a theory that about eighty to a hundred million years ago there was a world-wide change of climatic conditions, involving the development of greater temperature extremes through the day and the year. Faced with that, the makeshift mechanisms of the large reptiles broke down, and earth's most ponderous and magnificent land creatures died out.

However, perhaps as early as a hundred million years before that climatic change, when the age of the dinosaurs was in its first stages, some primitive reptiles had developed chemical devices that maintained even body temperatures, whatever (within reason) the temperature outside. These became the first of the class *Mammalia,* to which we ourselves belong. Later still, by twenty to fifty million years, a second group of reptiles made the same development and became the class of *Aves* (the Latin word for "birds").

In both classes, the body temperature was maintained considerably higher than the usual temperature of the environment. (A higher temperature hastens chemical reactions and allows creatures to live more intensely and more quickly—the advantage of the hare over the tortoise, and never mind the fable.) To maintain a high temperature economically, however, it was necessary to cut down the rate of heat loss to the atmosphere. Otherwise, the process grew too expensive in terms of energy consumed.

The heat loss was cut down by keeping a layer of relatively motionless air next to the body, still air being one of the best heat insulators. The birds accomplished this by trapping air among a set of modified scales called feathers, mammals by trapping it among a set of modified scales called hairs. (The feathers are the more efficient of the two, by the way.)

The birds took to the air, as had the insects and certain reptiles—now extinct—before them. In doing so, however, birds found that the aerodynamic facts of life limited their size drastically. Thanks to their efficient breathing system (involving lungs that pump air instead of, as in insects, small tubes through which air must percolate) they could grow considerably larger than insects, but they still had to remain considerably smaller than many reptiles and mammals of the land. Flight also involved the thorough commitment of one pair of limbs to the formation of wings—beautiful for the job but for nothing else.

Some birds were able to attain great bulk (for example, the ostrich and even larger, but now extinct, relatives) by abandoning flight. This, however, did not turn wings back into legs (evolution never seems to work backward in this fashion). Instead, they simply lost their wings, more or less, and were left bipeds still.

So the mammals, with all four limbs relatively uncommitted and with the potentiality of large size, proved, on the whole, to be more advanced than the birds.

The possession of soft hair among the mammals, in place of the horny scales of the reptiles, offered a new advantage. The mammals were exposing a soft skin to the environment, a surface sensitive to all kinds of impressions, to which the reptilian armor was perforce insensitive. Sensitivity to the environment is one method of insuring more efficient control, another way in which mammals are more advanced than reptiles.

In fact, armor has always seemed to be a failure. Any attempt to imitate the mollusk is fatal or nearly so. The armored placoderms gave way to the much less armored sharks and bony fish. Armored amphibians (of which many existed in bygone ages) gave way to naked frogs and toads. Superbly armored reptiles of the past lost out to less-armored reptiles of the present; and of present reptiles the better-armored turtles and alligators are not doing so well as the less-armored lizards and snakes.

Many early mammals developed armor and became extinct; the modern armadillos and pangolins, which are mammals that still possess armor, are comparatively low in the scale of mammalian development and are relatively unsuccessful.

Temperature control did one more thing for mammals (and birds, too). It made necessary the development of extended child care. Or, if you care to be more dramatic, warm blood led to mother love.

Temperature control, you see, is more easily maintained in a large organism than in a small one. All parts of the mass of an organism produce heat, but heat is lost only at the surface. A small creature has more surface per unit volume, hence loses heat at a greater rate. That is one reason, incidentally, why arctic mammals tend to be larger than mammals of warm climates. Large animals withstand cold better.

All this means that the most critical time in the life of a mammal, as far as heat control is concerned, is when it is smallest, when it is young or, most of all, when it is an embryo.

The eggs of a bird are much like those of a reptile, but the bird embryo is more dependent on equable warmth than the reptile embryo is. Those species of birds survived best which were most efficient in building nests, incubating eggs to the hatching stage, and feeding the young—all at considerable inconvenience to the adults.

The mammals go even further, but in stages.

The class *Mammalia* is divided into three subclasses. The first, *Prototheria* (Greek for "first beasts"), is still represented by two living species, one of which is the duckbill, or platypus. This subclass still shows many reptilian characteristics; its members are imperfectly warm-blooded, but they have hair, which no true reptile has, and they produce milk by means of mammary glands (whence the name "Mammalia"), which no true reptile does.

The Prototheria lay eggs, as reptiles do, but the embryo has progressed quite far in its development by the time the egg is laid, so that the incubation period, with all its special dangers, is relatively short.

The next subclass of Mammalia is *Metatheria* ("mid-beasts"), which includes such marsupials as opossums and kangaroos. Here another step is taken. The laying of the egg is so long delayed that it hatches before emerging from the body of the mother. It is an embryo at an early stage that actually emerges, but these embryos have just enough strength to make their way to the mammary glands of the mother, which are usually enclosed in a special pouch. In this pouch the young complete their development.

So far, the mammals have not really advanced spectacularly over birds or reptiles. Even the reptiles, in some cases, take a rudimentary care of their young, and there are some reptilian species that bring forth their young alive in a manner similar to that of the Metatheria (though without the pouch for postnatal care).

The Prototheria and Metatheria were the only mammals existing during the first fifty to a hundred million years of mammalian development; during the great age of the reptiles, they were only scurrying creatures that could scarcely lay claim to world mastery in the presence of the dinosaurs. But then climatic changes killed the largest of the reptiles without seriously endangering the warm-blooded mammals, and at about the same time the third and last subclass of the Mammalia was developed.

These were the *Eutheria* ("true beasts"), which underwent the crucial development of a special organ called the placenta, through which the developing embryo could absorb food and oxygen from the mother's circulatory system and into which it could discharge wastes. This makes longer gestation periods possible (some, as in the case of elephants or whales, are two years long). A long gestation period makes it possible for young to be born at advanced stages of development so that infant mortality is cut down to unprecedentedly low levels.

Thus the *placental mammals* (as the Eutherians are commonly called) have most nearly conquered the vicissitudes of the environment and are the most advanced group on the earth. Not only did they establish a clear ascendancy over the remaining reptiles, but they completely replaced the prototherians and metatherians everywhere but in Australia, which had broken away from the Asian mainland before the placental mammals had developed. (There are a couple of species of opossums in the Americas, as the only exceptions.)

But not all placental mammals are equally advanced. One thing that marks them apart from other groups of animals is the superior development of the brain, and in this they vary. Superior brain development, generally, is probably the consequence of life on dry land (which allowed the development of appendages), of soft, exposed skin, and the consequent development of improved sense organs. All these multiplied the stimuli to which the mammals were exposed and the variety of response of which they were capable.

Naturally, then, the more efficient the sense organs and the more generally responsive the appendages, the more intense would be the development of the brain. For appendages to remain generally responsive, however, they must not become too specialized.

I have mentioned, for instance, the wing of the bird. It is a fast-moving appendage, admirably adapted to its purpose but good for nothing else. Similarly, the marvellously organized legs of horses, deer, and antelope are excellent devices for outracing the enemy, but are no longer useful for anything else.

On the other hand, raccoons and bears walk flat-footedly on their heel in primitive fashion (as we ourselves do), and their paws can be used for a variety of tasks other than supporting the body. The members of the dog family, and some of the rodents also, retain the ability to use their paws as exploring devices. The elephant has developed a trunk that is the nearest thing any land creature can exhibit comparable to the tentacle of a squid.

The generally useful appendage reaches a climax in the order of *Primates,* a relatively primitive group of mammals, the ancestors of which took to the trees perhaps fifty million years ago. Life in the trees made it necessary, apparently, to develop hands that could grasp and long, flat-nailed fingers that could manipulate. The sense of sight was sharpened, and the eyes were brought about to the front of the head for binocular vision.

In the more advanced genera of the primates, particularly among the tailless apes (including man), one of the fingers, the thumb, is well developed and faces the other four, converting the hand into a still better grasping device.

The primates are, not surprisingly, the most intelligent of the mammals, and man, with the best-developed hand, is, not surprisingly, the most intelligent of the primates.

By using his brains and hands, man was able to extend the two most fundamental developments of land life generally. This was accomplished before modern man appeared on the scene. (The first manlike creatures were developed more than half a million years ago, but *Homo sapiens* is only twenty thousand years old.) Man learned to control fire and make use of clothing, thus adding to the efficiency of warm-bloodedness. Man also developed the systematic use of tools, which equipped him with artificial, fast-moving appendages, each of which might be thoroughly specialized. This gained the advantages of specialization without the loss of basic non-specialization.

And so, almost inevitably, chordate development (which is schematically summarized in the accompanying table) ends in *Homo sapiens* as lord of the earth, with

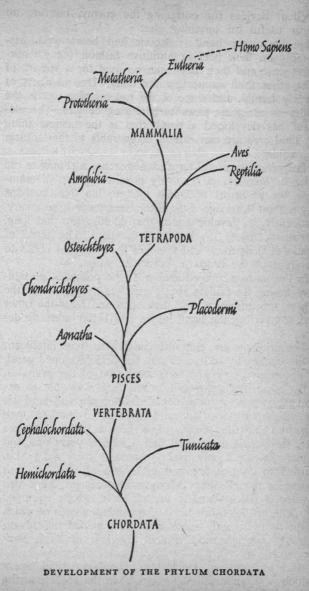

DEVELOPMENT OF THE PHYLUM CHORDATA

no enemy capable of facing him—except, of course, man himself.

So, you see, by the middle of the nineteenth century, it was possible to answer the question, "Where do babies come from?" by tracing matters back to an ancient cell in a primeval ocean.

William Schwenck Gilbert, the famous English comic-opera writer, said it well, although he said it jokingly. In his *The Mikado*, first produced in 1885, twenty-six years after *The Origin of Species*, Gilbert has Pooh-Bah (the satirical picture of an unprincipled aristocrat) say:

I am, in point of fact, a particularly haughty and exclusive person, of pre-Adamite ancestral descent. You will understand this when I tell you that I can trace my ancestry back to a protoplasmal primordial atomic globule. Consequently, my family pride is something inconceivable.

8

The Multiplication
of Cells

Of course, the question, "Where do babies come from?" does not usually refer to the ultimate origin of life or even of the species. The questioner is interested in the immediate origin of the individual baby, and is satisfied with an answer that goes no further back in time than is necessary to reach the sexual activity of the parents.

And yet the manner in which the sex act resulted in childbirth was not even vaguely understood until modern times. In prehistoric times, as mentioned in Chapter 1, no connection between the sex act and childbirth was recognized in many cases. Intercourse was valued for its own sake; as for children, that was entirely the mother's concern.

When the role of fatherhood was understood, however, the pendulum swung in the other direction. Man in general adopted an entirely proprietary air toward children. After all, when the male implanted his seminal discharge within the vagina of the female, he was much like a farmer sowing seed (the word "semen" comes from a Latin word meaning "to sow"). The lady in the case was reduced to a mere passive carrier of his child; she merely supplied the ground, so to speak, in which the male seed came to fruition. When a couple proved childless, the woman, like ground that would not bear, was said to be barren.

This view was pleasing to male vanity, and the first microscopic discoveries in connection with reproduction seemed to bear it out.

In 1677, a young man named Johann Ham working for Van Leeuwenhoek (mentioned in Chapter 1 as the first to observe microscopic forms of life) decided to see what semen looked like under one of his master's microscopes. He found, probably to his surprise, that it was full of wriggling little objects, for all the world like tiny tadpoles. He called them "animalculae seminis," which is Latin for "little animals in the seed." Van Leeuwenhoek took over and described them fully in his letters to England's Royal Society. We now call the objects spermatozoa (singular, *spermatozoon*) which means "animal seed" in Greek.

Did this not seem to bear out the importance of semen and the male? There were the little seeds waiting to be implanted in the female soil. Later microscopists, less conscientious or objective than Van Leeuwenhoek (who never reported anything proved wrong by later scientists with better instruments), let their convictions sway their judgment. They were sure that they could see tiny human beings *("homunculi")* already formed within human spermatozoa and drew fanciful pitcures showing this.

Although better microscopes later refuted this, the sperm supporters maintained their strength for many years.

It was not until 1827, in fact, that a corresponding structure was found within the female mammal. (Female birds and reptiles, to say nothing of fish and insects, laid eggs; but female mammals showed no signs of such an intimate connection with childbirth.) The German naturalist Karl Ernst von Baer first found tiny eggs or *ova* (singular, *ovum*, the Latin word for "egg") within the female. She, too, now had her version of a "seed."

Once the cell theory was established, the ova and spermatozoa were recognized for what they were. Schwann, one of the founders of the cell theory, himself stated, in 1838, that the ovum was a cell, a single cell. In 1841, the Swiss biologist Rudolf Albert von Kölliker suggested that the spermatozoon, as well, was a cell. No one has argued with either viewpoint since, and ova and spermatozoa are now frequently referred to as *egg cells* and *sperm cells,* respectively (they can be lumped together as *sex cells*).

It was soon realized (particularly from the study of the pollination of plants) that for any individual creature to develop, a sperm cell had to fuse with an egg cell. The combination of the two formed a *fertilized ovum,* which then proceeded to divide and subdivide. There remained some question as to whether one or more than one sperm cell was required for fertilization. Sperm cells are so small, you see, and many millions exist in a single seminal discharge. This was settled in 1879 by a Swiss zoologist, Herman Fol, who, working with starfish, actually witnessed the entrance of a single sperm into an ovum and consequent fertilization.

Just knowing that sperm cells and egg cells both existed, and that a combination of the two was required for development and birth, did not help in settling the problem of which one was the predominant factor. There were proponents for each point of view, with those in favor of sperm having somewhat the stronger case, at least in appearance. There was no doubt that the sperm cell was an active little thing with a lashing tail and a full measure of vitality, while the egg cell was only a passive blob.

To be sure, the egg cell was thousands of times as massive as the sperm cell. The mammalian egg cell is about the size of a pinhead and is the largest cell in the body, whereas the sperm cell is much smaller even than the average cell. And yet, the superior size of the egg cell was due only to the presence of food. (The eggs of birds and reptiles, which are larger still—and yet are single cells, even up to and including the ostrich egg—are larger only by virtue of a still greater food supply.) Could it be that the egg cell was food entirely and that it was the sperm cell which alone added the spark of life?

Actually, it was Mendel who settled this problem. His crossings of the pea plants made it clear that there was no difference whether the pollen of variety *A* fertilized the ovules of variety *B*, or the pollen of variety *B* fertilized the ovules of variety *A*. Consequently, the conclusion was that male and female contributed equally to the development of the child. For all their difference in size and motility, egg cell and sperm cell contributed matching shares to the fertilized ovum.

Naturally, what Mendel discovered had to wait until Mendel himself was discovered, and so the scientific world, generally, did not reach an absolute solution to the egg/sperm controversy during the nineteenth century.

Meanwhile, however, the study of how the single-celled, fertilized ovum multiplied itself into uncounted millions of cells, developing through various embryonic stages to the point where an organism was ready to break the egg-shell (or emerge from the womb) and take up an independent existence, itself developed apace and became the science of *embryology*.

The first modern embryologist was Von Baer, the man who had discovered the mammalian ovum. He also first described the manner in which the fertilized ovum, as it developed, formed three generalized cell layers out of which the organs developed. These he referred to as *germ layers* ("germ" in its older sense of anything small in which life exists and out of which more complicated life can develop). It was the German physician Robert Remak who in 1845 gave the germ layers their modern names of ectoderm, endoderm and mesoderm.

Embryos were found to develop odd features in the course of their growth. For instance, in 1829, a German anatomist, Martin Heinrich Rathke, discovered that at one stage in their development the embryos of birds and mammals formed gills, which, however, did not persist.

By the time Darwin's *Origin of Species* was published, there was a whole collection of embryological observations which showed that embryos developed, and then dropped, a number of organs which they did not need and never used.

With Darwin's light blazing into every corner of biology, the thought came at once: is the embryo harking back to its evolutionary past? Zoologists began enthusiastically investigating more primitive forms of life in search of more evidence of this nature.

For instance, in 1827, Von Baer had discovered that at one stage in its development, the mammalian embryo possessed a cartilaginous rod down its back. In 1848, the British anatomist Sir Richard Owen called it a notochord. At the time, the notochord was not known outside embryos.

But then, between 1866 and 1871, the Russian zoologist Alexander Onufrievich Kovalevski studied the amphioxus intensively. It had a notochord running the length of its body throughout its life, and its development was similar to the early development of vertebrate embryos. The amphioxus was more primitive than the vertebrates, then, but related to them, and the embryo, in developing a notochord (later to be replaced by vertebrae), was harking back to its own prevertebrate ancestors.

In fact, an elaborate scheme could be built up with regard to embryonic development. For instance, the human being begins life as a fertilized ovum, a single cell resembling, in that respect, the one-celled creatures that swarmed in the primordial ocean.

The ovum divides and redivides, forming a group of cells that cling together and are called a *morula* (Latin for "little mulberry"). If, after the first one or two cleavages, the resulting cells separate through some accident, each of the separated cells can give rise to a complete and separate organism. In this way, identical twins, triplets, quintuplets even, can arise. This is evidence that before it becomes a true multicellular organism, the developing ovum passes through a brief stage as a cell colony.

As cleavage continues, the multiplying cells arrange themselves to form a hollow sphere (something like the Volvox) which is called the *blastula* (Latin for "little bud").

The blastula then sucks inward at one point to form a cup that is very similar to the coelenterate body plan. An ectoderm on the outside and an endoderm on the inside have appeared. This is the *gastrula* (Latin for "little gut"). The mesoderm forms out of the endoderm, as in the echinoderms, and then the various organs appear.

The notochord appears, as in the primitive chordates, and is replaced by vertebrae. Gills are developed as in fish and are replaced by lungs. Hair and a tail develop, as in the lower mammals, and are then lost. Even the bones are first cartilaginous, as in the sharks, and only slowly become bony, that part of development concluding well after birth.

The man who first accepted this view of embryology and riveted it into prominence by means of a memorable phrase was Ernst Haeckel, the foremost proponent of Darwinism in Germany. In about 1874, he used a phrase which can be translated into three polysyllabics: "Ontogeny recapitulates phylogeny."

"Ontogeny" is the development of an individual organism and "phylogeny" is the development of the group to which that organism belongs. What the phrase means, then, is that every human being, for instance, must go through all the stages of development through which his ancient ancestors went. The recapitulation might be hasty and slurred, since a billion or more years of development must be raced through in nine months, but it is there.

Biologists these days do not accept Haeckel's thesis completely, but it remains a rather attractive simplification.

Just as each tadpole, today, must emerge from water and personally conquer land as did his ancestors hundreds of millions of years ago, so must each human embryo make the same conquest in memory of the same grand event. Each embryo must develop gills first and only thereafter lungs. But this is done so hastily that when the child emerges into the outside world with the transition safely made and the newly expanded lungs pumping air in a vigorous squall, no proud parents would ever guess that their newborn darling had been more fishlike than manlike for a small space of time.

Consequently, even if the question, "Where do babies come from?" is restricted to the individual organism, the answer must, one way or another, hark back to Pooh-Bah's "protoplasmal primordial atomic globule."

Since all life seemed to begin with the cell, the key to a better understanding of life apparently had to rest within the cell. How did cells manage to multiply? How, in doing so, did a skin cell always divide into two skin cells, and never into two liver cells, while a liver cell always gave rise to more liver cells and not to skin cells? For that matter, why did the fertilized ovum of a giraffe always turn into a giraffe, while the very similar-appearing fertilized ovum of a human being always turned into a human being?

The answer had to lie within the cell, and scientists were trying to peer within it even before the cell theory was announced and the importance of the cell recognized.

As early as 1781, the Italian naturalist Felice Fontana made out some detail in skin cells contained in the slime from an eel, which he was studying under the microscope. Fontana reported that within each cell he saw a still smaller body with an appearance different from the remainder of the cell. This still smaller body is now called the cell *nucleus* (plural, *nuclei*) from a Latin word meaning "little nut." It has been since discovered that all cells contain them. (Apparent exceptions are the red blood cells of man and most other mammals. They do not contain nuclei. But then, for that very reason, they are not considered true cells and are usually referred to as "red corpuscles," the word "corpuscle" coming from a Latin word meaning "little body.")

Just as the cell is marked off from the rest of the world by a thin cell membrane, so the nucleus is marked off from the rest of the cell by an even thinner *nuclear membrane*. The protoplasm within the cell but outside the nucleus eventually received the name of *cytoplasm*.

Inside both nucleus and cytoplasm were all sorts of grainy materials that could not be clearly made out and that must have tantalized the cytologists no end, as their frustrated peering gave them headaches, both figuratively and literally.

The turning point came in the middle of the century when it was discovered that the molecules of certain dyes would add to various chemicals within the cell and change them from dull shades of gray to what would now be called "living Technicolor."

The first dyes studied did not show much differentiation in their behavior. For instance, in 1850, the dye carmine was found to color the contents of the cell a reddish-purple—the first definite proof that there was actually material within the cell. Another type of dye fastened onto the nucleus particularly, changing it to a small, blackened area within a largely colorless cytoplasm. The nucleus became easier to see.

Of course, stains killed the cell, so that cytologists found themselves looking at dead material rather than living protoplasm. However, a dead cell with contents that can be seen has an advantage over a living cell with almost invisible contents, and cytologists went over to the use of stains with great delight.

It was just about this time, too, that chemists were learning how to make hundreds of artificial dyes, and this meant cytologists had many colored substances from which to choose. These new dyes did sometimes differentiate among the cell structures. Some were attracted to one part of the cell contents and some to another. In this way, it was discovered that cells contained tiny structures of definite shape. These tiny structures were called *organelles* ("little organs") or, more vaguely, *particulates* ("little particles").

In particular, the German cytologist Walther Flemming discovered in 1879 that various basic aniline dyes strongly stained certain portions of the substance within the nucleus, leaving the rest of the nucleus and all of the cytoplasm untouched. These stainable portions within the nucleus he called *chromatin,* from a Greek word for "color."

Using this dye, Flemming found he could follow events during the process of cell division. Various cytologists had in previous decades reported what they could see of those events, but that had not been much, and most of what was described turned out, in the light of later knowledge, to have been wrong.

But Flemming, with his aniline dyes, could now see details more clearly than anyone ever had before. To be sure, the dye killed the cell and trapped it at a single moment of cell division. However, Flemming used slices of growing tissue, in which there were individual cells at all stages of division. He would catch various cells in "stills" representing each of the different stages. Putting them all together deftly, he could (and did) reconstitute the "motion picture" of cell division.

His analysis appeared in 1882, in an epoch-making book, the title of which (in English translation) was *Cell-Substance, Nucleus, and Cell-Division.*

The various stages of cell division are called "phases" (from a Greek word meaning "appearance," since, after all, it is by changes in the appearance of the stained cell that we distinguish the stages).

To begin with, there is the "resting cell," the cell that is growing and is not yet ready to divide. This is *interphase* ("in-between appearance"). The nucleus is then a small ovoid within the cell, well marked off by its membrane, while the chromatin material is present in small splotches within the nucleus. Just outside the nucleus is a tiny body now called the *centrosome* (from Greek words meaning "central body"), a name first given it in 1888 by the German physician Theodor Boveri, because of its central role in cell division. In the middle of the centrosome is a fine dot called the *centriole* ("little center"), and radiating from the centriole are fine lines called *centromeres* ("parts of the center"). Centriole and centromeres together look rather like the conventional drawing of a star, and Flemming's name for them was *aster* (the Greek word for "star"), which is still sometimes used.

When the time for cell division draws near, changes begin to take place; these early stages are called *prophase* ("first appearance"). The centriole splits in two, and the two centrioles that result begin to move apart, connected by some of the centromeres. The chromatin material in the nucleus begins to collect together into wormlike threads. Finally, the nuclear membrane begins to break up, and the material within the nucleus mixes with the cytoplasm.

(The threads of chromatin material that begin forming during prophase were named *chromosomes*—from Greek words meaning "colored bodies"—by the German cytologist Wilhelm Gottfried Waldeyer, in 1888. This name has become so popular that chromatin material is almost universally spoken of as chromosomes whether it is condensed to the threadlike appearance or not.)

The next general stage is *metaphase* ("middle appearance"). The two centrioles are on opposite sides of the cell, with centromeres still connecting. The chromosomes have collected at the middle of the cell between the centrosomes.

Next is *anaphase* ("further appearance"), where the fine lines of the centromeres break and each of the resulting halves contracts toward the centriole nearest it. The chromosomes, as though entangled in the centromeres, are dragged in either direction. One half go toward one side of the cell; the remainder toward the other. As the center of the cell is cleared of chromosomes, the cell begins to pinch inward about its equator.

Then in the *telophase* ("final appearance"), the pinching is completed and, as the two daughter cells form, a new nuclear membrane forms in each, enclosing the chromosomes, which begin relaxing once more into chromatin material.

Then comes a new interphase, except that now we have two cells where one cell existed before. That whole process was called *mitosis* by Flemming from a Greek word

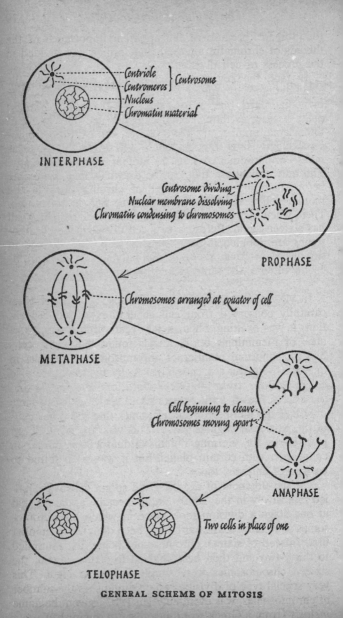

GENERAL SCHEME OF MITOSIS

for "thread" because of the apparent importance of the threads of chromatin material that formed and divided in the process of cell division.

In fact, the key point in the entire process was the clean-cut and careful division of the chromosomes into two equal parts, one set for each daughter cell. This might make it seem as though each daughter cell only had half the original number of chromosomes. But careful chromosome counts show that daughter cells invariably have as many chromosomes each as the mother cell had originally. The only conclusion that can be drawn is that, sometime before metaphase, the number of chromosomes in the mother cell is doubled, from sixteen, say, to thirty-two. Then, when the chromosomes pull apart, each daughter cell has sixteen again. Before a new cell division, the chromosomes are doubled in each daughter cell, then divided in two again. Alternate doubling and halving keep the chromosome number the same generation after generation.

In view of the fact that the chromosome number is so carefully conserved from cell generation to cell generation, it was exciting when, a few years after the appearance of Flemming's book, it was found that some cells had an abnormal number of chromosomes. The cells in question were the sex cells (which, by the way, are also called *gametes,* from the Greek word for "marriage") and these were found, both egg cells and sperm cells, to have only half the number of chromosomes that the other cells of the particular organism had. This was first pointed out in 1888 by the German botanist Eduard Strassburger in connection with certain plants, but it was soon found to be true of animals, too.

The development of egg cells and sperm cells was studied intensively in the 1890's and it became apparent that at one stage a division took place in which the chromosomes were halved without having been doubled in number first. Intsead of sixteen chromosomes being first mutiplied to thirty-two and then being halved to sixteen again, the sixteen chromosomes were simply halved to eight. This very special type of division, which diminishes the number of chromosomes, was eventually (in 1905) given the name *meiosis* (from a Greek word meaning "to diminish").

The sex cells are, therefore, said to be *haploid* ("single form" in Greek) and the remaining body cells *diploid* ("double form").

The haploidy of the gametes obviously is for a good reason. Eventually, a sperm cell enters an egg cell and the two nuclei fuse. Each contains a half portion of chromosomes, but once they come together in the fertilized ovum, the full number is restored. The fertilized ovum is a diploid cell and gives rise to all the diploid cells of the body by ordinary mitosis and, eventually, to haploid gametes of the new generation by meiosis.

By 1900, then, the adventures of the chromosomes during cell division were well documented, but no one had quite figured out what it all meant. In that year, Mendel's theories of heredity were shouted forth to men of science and, almost at once, it was noticed that Mendel's hereditary factors behaved exactly as the chromosomes behaved.

The first scientist to see clearly how the two lines of research fused together into a neat unit and to translate Mendel's findings into chromosomal terms was the American biologist W. S. Sutton. His ideas were presented in 1902.

Sutton pointed out that one could suppose that the chromosomes occurred within the nucleus in pairs. In other words, a cell with sixteen chromosomes would have eight pairs of chromosomes (which we ourselves can think of, for the sake of clarity, as A and A', B and B', C and C', and so on).

If the chromosomes carried the hereditary factors, then for every particular characteristic there would be two such factors, one on a chromosome and one on its pair.

In the formation of the gametes, each sperm cell and egg cell ended up with half the usual number of chromosomes; but not just any half at random. They ended with precisely one chromosome of each pair, but which of the pair is purely a matter of random happening. For instance, a particular gamete might get chromosomes, A, B, C, D, A', B', C', D', A, B', C', D, or A', B, C', D, but it would never get A, A', B, B', with no C or D at all.

If, for a particular characteristic, an organism carries two different alleles of the gene involved, one (governing red flowers, for instance) will occur on a particular chromosome, say, *A*, while the other (governing white flowers, perhaps) will occur on that chromosome's pair, *A'*. In the formation of the gametes, half of those formed will get a chromosome combination that will include *A*, and the other half a combination that will include *A'* (it being a matter of chance, like tossing heads or tails on a coin). Half the gametes, therefore, will end up with the chromosome carrying the red-flower gene, and the other half with the chromosome carrying the white-flower gene.

This is true for both sperm cells and egg cells, and the combination of the two will form a chromosome pair that will contain either *A* and *A*, *A* and *A'*, *A'* and *A*, or *A'* and *A'*. In terms of genes that would be *RR*, *RW*, *WR*, and *WW*.

Since chromosomes are of identical nature, whether contained in egg cell or sperm cell, and since each type of gamete carries the same number, one of each pair, egg cell and sperm cell make equal contributions to inheritance. The disparity in size between the two gametes arises from the fact that the sperm cell is little more than a tightly packed bag of chromosomes, with a tail added for motility. The egg cell is much larger because it carries a food supply to tide the developing embryo over until a placenta has been formed.

If chromosomes *A* and *A'* carry genes for a particular flower color, while chromosomes *B* and *B'* carry genes for a particular seed color and chromosomes *C* and *C'* carry genes for a particular length of stem, all these characteristics will show up in the new generation according to the random distribution of chromosomes. A particular sperm cell may get *A*, *B'* and *C*, while a particular egg cell may get *A*, *B* and *C'*. The combination will then be *AA*, *BB'* and *CC'*. Each generation involves a random shuffling of each pair of chromosomes.

If you were to look back to Mendel's experiments now, you would see how easily all his results can be understood in the light of the behavior of chromosomes.

9

The Changing
Chromosomes

Once the connection was made between chromosomes and heredity, there was a powerful urge to study the behavior of chromosomes more intensively and try to match that behavior with the facts of inheritance even more closely. The natural tendency would be to study human beings, since it is in *Homo sapiens* that we are most interested. However, individual human beings have children no oftener than once a year and then, usually, but one at a time.

Besides this, each human cell contains forty-six chromosomes (twenty-three in each gamete) and these are too many to handle easily. In so complicated a fashion do these chromosomes intertwine themselves in the cell that it was not until 1957 that the correct number was finally determined. Until then, the general impression had been that the human cell contained forty-eight chromosomes.

What was needed was a simpler type of organism: one that was small and with few needs, so that it might easily be kept in quantity; one that bred frequently and copiously; and one that had cells with but a few chromosomes. An organism which met all these needs ideally was first used in 1906 by the American zoologist Thomas Hunt Morgan. This was the common fruit fly, of which the scientific name is the much more formidable *Drosophila melanogaster* ("the black-bellied moisture-lover"). These are tiny things, only about one twenty-fifth of an inch long, and can be kept in bottles with virtually no trouble. They can breed every two weeks, laying numerous eggs each time. Their cells have only eight chromosomes apiece (with four in the gametes).

More genetic experiments have been conducted with *Drosophila* in the past half-century than with any other organism, and Morgan received the Nobel prize in medicine and physiology in 1933 for the work he did with the little insect. Enough work was done with other organisms, from germs to mammals, to show that the results obtained from *Drosophila* studies are quite general, applying to all species.

One of the first things to be noticed about the chromosomes of *Drosophila* was that they did, indeed, seem to occur in pairs, as the Sutton theory had stated. Thus, of the eight chromosomes in the *Drosophila* cell, two were tiny, dotlike affairs; two were short, straight lines; and four were longer and bent at the middle into V-shapes. In the formation of a gamete, the four chromosomes it ended with consisted of one dot, one straight line and two V-shapes. Gametes were not observed to have sets of four chromosomes that included both dots, or both straight lines, or three V-shapes. There seemed a straight and unquestioned separation of pairs.

To be precise, though, it was in the female *Drosophila* that the pairs were exactly alike in appearance. In the cells of the male *Drosophila*, there were two dots and four V-shapes, to be sure, but the remaining chromosome pair was not a perfect match. One of this pair was a straight line as in the female, but the other was a bit longer, with the added portion bent into a J-shape.

Something of this sort had been observed in 1905 by the American zoologist Edmund B. Wilson, who was studying bees. He was the first to use the expression X-chromosomes for the special, not necessarily matched pair of chromosomes, and the name stuck. Thus, the female *Drosophila*, with two straight-line chromosomes, is said to have two X-chromosomes. The male *Drosophila*, with one straight-line and one J-shape, is said to have one X-chromosome and one Y-chromosome.

When a female *Drosophila* forms egg cells, each egg cell gets one of each chromosome pair, which means that each gets one X-chromosome. When a male *Drosophila* forms sperm cells, each sperm cell likewise gets one of each chromosome pair. But in the case of the male there is a mismatched pair, which makes for the formation of two types of sperm cells. Half the sperm cells, as their share of the division, end up with the X-chromosome, the other half with the Y-chromosome. Suppose we call the first group of sperm cells X-sperm, and the second, Y-sperm.

During the fertilization of a given egg cell (which always contains an X-chromosome, remember), one of two alternatives will take place: either an X-sperm will consummate the fertilization, or a Y-sperm will. In the former case, the fertilized ovum will contain two X-chromosomes and will give rise to a female. In the latter case, the fertilized ovum will contain an X-chromosome and a Y-chromosome and give rise to a male. Whether it is an X-sperm or a Y-sperm that does the fertilizing is purely a matter of chance, and since there are equal numbers of both varieties of sperm, the result is that, in the long run, male and female offspring are born in virtually a one-to-one division.

The details of sex determination vary from species to species, but in most cases it is essentially similar to that found in *Drosophila*. In the case of human beings, the Y-chromosome is a little blob as compared with the X-chromosome, which is of normal size, but in men, as in fruit flies, the female of the species owns two X-chromosomes, the male an X-chromosome and a Y-chromosome.

In some species, there is no Y-chromosome at all, the X-chromosome in the male simply being unpaired. In other species (some birds, for instance) it is the female which has a mismatched pair of chromosomes, rather than the male.

And yet, despite the way in which chromosomes matched Mendel's findings and the manner in which, almost at once, they explained sex determination at the moment of fertilization, chromosomes alone were obviously insufficient to explain the facts of heredity. The individual chromosome could not possibly govern one particular characteristic and no others, since there are far more inheritable characteristics in any given organism than there are chromosomes. It is inevitable to suppose (and Sutton, when he first worked out the chromosome theory, suggested this) that each chromosome controls a variety of characteristics; that the chromosome, in other words, is made up of a great number of genes.

If the chromosome does indeed contain many genes, then all those genes must move together as the chromosome moves. If chromosome A is to be found in a particular gamete, it must contain genes A1, A2, A3, A4, and so on. If the pair of that chromosome, chromosome A', is found in the gamete instead, it must contain genes A1', A2', A3', A4', and so on. Provided the chromosome itself always remained intact and a unit, gene A1 would always be accompanied by gene A2, and never by A2'. A gamete would either have both A1 and A2 (along with all the other genes on that chromosome) or neither.

This transfer of a series of genes, as a linked unit, from generation to generation is referred to as *gene linkage*.

When Mendel studied the seven varieties of sweet peas, he chose characteristics (by a peculiar and most fortunate coincidence) that happened each to be on a separate chromosome. Each could therefore end up in a gamete, as part of its own chromosome, regardless of whether any other particular chromosome made it or not. Thus, the seven characteristics varied independently of one another, and Mendel was able to work out his "law of independent assortment."

Had he just happened to pick on two or more characteristics, with genes occupying the same chromosome, he would have found that those two characteristics were almost always both present in a particular plant, or both absent. This would have complicated his results and delayed the working out of his theories.

We can take an example of linkage in *Drosophila*. This insect carries a gene that controls eye color and one that controls wing color, and both are on the same chromosome. Among the alleles of the eye-color gene are one that produces red eyes and one that produces white eyes, with the red-eye allele dominant. Among the alleles of the wing-color genes are one that produces yellow wings and one that produces gray wings, with the yellow-wing allele dominant.

Suppose we call the eye-color alleles RE (red eye) and WE (white eye), and the wing-color alleles YW (yellow wing) and GW (gray wing).

Next suppose that a male *Drosophila* with a chromosome pair which both carry RE and YW is mated with a female *Drosophila* with a chromosome pair which both carry WE and GW. All the sperm cells produced by the male have the RE/YW chromosome. All the egg cells produced by the female have the WE/GW chromosome. All the fertilized ova contain one of each and all the new generation of *Drosophilae* have red eyes and yellow wings, like the father. The mother's characteristics of white eyes and gray wings seem to disappear.

Now suppose that these hybrid *Drosophilae* are bred among themselves. Each of them contains a chromosome pair of which one is RE/YW and the other WE/GW. Consequently, half the egg cells (or sperm cells) will contain only the RE/YW chromosome, and half the WE/GW chromosome. There will be four possible combinations: an RE/YW sperm cell with an RE/YW egg cell; an RE/YW sperm cell with a WE/GW egg cell; a WE/GW sperm cell with an RE/YW egg cell; and a WE/GW sperm cell with a WE/GW egg cell.

In the first combination, the result is an individual with an RE/YW-RE/YW pair of chromosomes. It will be a red-eyed, yellow-winged individual which, mated with others like itself, will breed true. The next two combinations will result in individuals with an RE/YW-WE/GW pair of chromosomes. They will be red-eyed and yellow-winged also, but, mated among themselves, they will not breed true. Finally, the last combination will be an individual with a WE/GW-WE/GW pair of chromosomes and these will be white-eyed and gray-winged.

The important thing here, however, is that because of the presence of two genes on a single chromosome, whenever a descendant of the original *Drosophilae* progenitors has red eyes, it also has yellow wings (at least in the case used here as an example); and whenever it has white eyes it also has gray wings. The two characteristics are linked.

So much for theory. In actual practice, if you were to start by crossing a red-eyed, yellow-winged insect with a white-eyed, gray-winged one as described above, most of the descendants would have either red eyes plus yellow wings or white eyes plus gray wings. There would be a comparatively small number, however, that would have red eyes plus gray wings, or white eyes plus yellow wings.

How can such a thing happen? Well, remember that the notion of gene linkage depends on the assumption that the chromosome always remains intact. Close observation of the behavior of chromosomes during cell division proves that they do not necessarily remain intact. Each pair of chromosomes, as it lines up at the center of the cell during metaphase, is more or less intertwined. Every once in a while there would be an interchange of material before the separation took place. A smaller or larger section of the two chromosomes would change places so that chromosome A would suddenly find its top third to be what had originally been the top third of its pair, chromosome A'. In exchange, chromosome A' would find itself blended with the top third of chromosome A. This phenomenon is called *crossing over*.

Suppose that crossing over takes place during meiosis in a *Drosophila* that has an RE/YW chromosome and a WE/GW chromosome. The portion of the chromosomes that crosses over may contain the eye-color alleles but not the wing-color alleles. The result is that the new crossed-over chromosomes are WE/YW and RE/GW. If a WE/YW sperm produced by a *Drosophila* which has undergone such a crossover fertilizes an ordinary WE/GW egg cell, then the result is a WE/YW-WE/GW individual, one with white eyes and yellow wings. In this way, the red-eye plus yellow-wing linkage has been broken.

In 1911, Morgan pointed out that crossing over could be used to determine the positions of particular genes within a chromosome. Suppose, for instance, that there are just one hundred and one genes lying along a particular chromosome and that, in crossing over, the material that shifts from one chromosome to the other can break off at any of the hundred dividing lines between successive genes.

Now imagine two genes at opposite ends of the chromosome. No matter where along the chromosome the breaking point comes during crossover, those two opposite-end genes separate. Before crossover, they are on the same chromosome; after crossover, they are on opposite chromosomes. Always! Working backward, two genes that are invariably separated every time there is a crossover must be at the two ends of the chromosome.

Suppose, on the other hand, that the two genes are so positioned that there are only twenty-nine other genes between them. The break during the crossover may come at any of the thirty gene-separation points between them. A break at any of those thirty points would separate the two genes with which we are concerned and place them on opposite chromosomes. There are, however, seventy dividing points that lie on one side or the other of both genes and do not lie between them. A break at any of those seventy would shift either both genes or neither. In either case, the genes would remain on the same chromosome.

In this case, consequently, the genes will be separated after thirty per cent of the crossovers and not separated after seventy per cent of them.

The more frequently genes are separated as a result of crossover, the further apart those genes must have been located in the original chromosome. By studying many generations of vast numbers of *Drosophilae,* detailed chromosome maps of the location of various genes on the four chromosomes have been prepared (the first was prepared in 1917). In theory, the same could be done for any species, but with more chromosomes per cell and longer intervals between generations and fewer individuals per generation, the job quickly becomes impractically complex.

The number of possible combinations that result from chromosome shuffling from generation to generation can be enormous. Consider the human being, for instance, with forty-six chromosomes. A baby inherits twenty-three from his father and twenty-three from his mother. Let us begin with the first pair of chromosomes and call those of the Father A and A′, while those of the mother are *a* and *a′*.

The baby can end with a chromosome pair consisting of *Aa, Aá,* or *A′á*. There are thus four combinations possible in the case of the first chromosome pair only. The same can be said of the second pair of chromosomes, the third pair, and so on down to the twenty-third pair. The total number of combinations is, therefore, 4x4x4 x4x4x4x4x4x4x4x4x4x4x4x4x4x4x4x4x4x4x4, that is, twenty-three 4's multiplied together, a process which can be written, briefly, as 4^{23}. This product comes out to almost 100,000,000,000,000 (a hundred trillion).

If any two of the four chromosomes involved at each position, say A, A', a and $\acute{a}$, happened to have identical alleles for every gene down the line, then some of the combinations would be identical. If for instance A and A' were absolutely identical in compositon, then a chromosome combination that included A and a would be the same as one that included A' and a. However, the existence of absolutely identical chromosomes is extremely unlikely. We are safe in saying that all or virtually all the hundred trillion different chromosome combinations that are possible would yield individuals with at least slightly different combinations of physical characteristics.

Since crossing over is common, the chromosomes themselves change from generation to generation. Eventually, every possible gene combination may come out in one individual or another. In a species such as man, the number of gene combinations is so great as to reduce the mere hundred trillion combinations of chromosomes to insignificance. It is very unlikely (unthinkable, in fact) that the number of different gene combinations possible will be exhausted during man's probable length of existence in this universe, and it is extremely unlikely that any two human beings (barring identical twins) would ever share precisely the same gene combinations.

And all this is true, even if we limit ouselves to normal mitoses—which we need not do.

In Chapter 4, I pointed out that De Vries had worked out his mutation theory before he rediscovered Mendel's law. He did this as a result of observing a colony of American evening primroses in which some of the plants were radically different from the others.

After Sutton advanced the chromosome theory of inheritance, De Vries' mutated primroses were looked at from a fresh angle and it was quickly found that they contained twice as many chromosomes as did the cells of ordinary primroses.

Apparently, every once in a while something went wrong with the development of a fertilized ovum. The fertilized ovum, instead of doubling the chromosome number, then dividing it in two among two daughter cells, would merely double it. No cell division would succeed immediately. Then, after a time, there would be a second doubling, followed by a normal division, and the matter would continue normally thereafter. However, because of that first doubling followed by no division, every cell in the plant would have twice the normal number of chromosomes, and this characteristic would be passed on to its descendants.

With time, organisms of various species were detected with even more than twice the normal complement of chromosomes—with three times the number, four times, five times and so on. Such conditions are referred to as *polyploidy* (from Greek words meaning "manifold"). Cases of polyploidy invariably mean a sudden change in characteristics.

You may wonder why a mere doubling of the chromosomes should affect physical characteristics. After all, if there had been a red-flower allele and a white-flower allele before, let us say, then when chromosomes are doubled, you would have two red-flower alleles and two white-flower alleles. Why should that make a difference?

Apparently, genes are more than independent little hereditary factors strung out upon the chromosome string. Each gene is affected by its neighbors and, in the long run, by all the other genes.

You might compare the genes of a cell to the instruments in a symphony orchestra. Each instrument in an orchestra has its own part to play in the symphony. You might listen to the first instrument play its part in isolation, then to the second, then to the third, and so on. You might be able to learn a lot about the symphony in this way, just as geneticists learned a good deal about heredity studying genes one at a time, as Mendel did.

However, the true sound of the symphony can only be heard when all the instruments are playing in unison. Similarly, the over-all characteristic of the organism can be understood only as a product of all the genes working in unison.

The symphony sound can be changed by merely doubling all the instruments, even though no new type of instrument were added. This would be "instrument polyploidy." Even switching the relative positions of the instruments would affect the over-all sound.

This latter case is also true of genes. Sometimes crossing-over processes get tangled. A piece of chromosome from the end of chromosome A might get attached to chromosome A′ upside down, or it might find itself in the middle of chromosome A′. In either case, all the genes would still be there, but their positions would be different; they would have different neighbors. Such imperfect crossovers would also produce organisms with new sets of physical characteristics.

However, despite all this, the most important mutations are not the result of polyploidy or of imperfect crossovers. They are the result of the formation of brand-new alleles that had not before existed.

10

The Changing Gene

Mutations were detected constantly in *Drosophilae,* and often there was no corresponding visible change in the number or structure of chromosomes to account for it. The easiest explanation was to suppose that there was some defect in the process by which new chromosomes were formed, some defect too small to be seen under the microscope.

The new chromosomes formed before each cell division must generally be the exact image of the original, gene for gene. This exactness, however, is merely general and not invariable. Occasionally, an inexact duplicate of a gene is formed—a new allele, a "mutated gene."

The mutated gene would reproduce itself—for the most part accurately—in subsequent cell divisions. It may find its way into sex cells and, if it is dominant, it may show up in terms of a visible characteristic in the next generation. If it is recessive, it will not, for it will be drowned out by the normal allele of that gene. However, even then, if a similar allele occurs in more than one individual, or even if it occurs only once but is passed on to numerous descendants, there may eventually be a mating that will combine the mutated genes into a "double dose" so that the new characteristic will show itself.

Do such mutations occur often enough to affect evolution? The answer is: Yes, mutations are not rare. Work on individual characteristics shows that it is not unusual to have a particular gene mutate once every hundred thousand times (or less) that new genes are formed. This may sound like a rare occurrence, but considering the number of times that genes are duplicated in the lifetime of an individual and the total number of organisms in the world, it is not surprising to learn that some geneticists have estimated that perhaps one organism in every ten is a *mutant* in the sense that it possesses one or more freshly mutated genes.

Mutations can result in small and nearly unnoticeable changes, or in large and drastic ones. Most are for the worse. A mutated gene will result in some malfunction or other that will place its owner at a disadvantage as compared with its neighbors. This is not theory, but is a conclusion based upon the observation of actual mutations in *Drosophilae* and other species.

How, then, can mutations be expected to advance evolution (as described in Chapter 5) if they are so apt to be for the worse?

Occasionally, a mutation does help fit a creature more firmly into its environmental niche. When that happens, the mutation survives, and the gene allele that caused it replaces the hitherto normal allele.

Since it is precisely those few mutations which help fit the organism to its niche that survive, it is those which we witness. The large majority of mutations, which work harm, will vanish and remove themselves from our sight, or at best remain visible in a small minority of individuals. In this way, we get a biased view of the workings of mutation and, if we are careless, think of it as driving organisms up the evolutionary tree to better and higher things. Then we begin to think that this is too marvelous to be accounted for by random forces only, and seek a supernatural explanation where one is not necessary.

The mutation theory dates back to the 1880's, and its implications, as far as evolution was concerned, were easily seen even then (as they were described in Chapter 5). However, it was difficult to study mutations in a systematic way, as long as geneticists simply had to sit around and wait for them to happen.

The turning point, in that respect, came in 1923, when the first reports were published that mice exposed to X-rays produced abnormal offspring. Following that lead, the American geneticist Hermann Joseph Muller began, in 1927, to expose *Drosophilae* to X-rays. Sure enough, the rate of mutation shot up. The type of mutations did not change; they were the same as those that resulted in the absence of X-rays. There were more of them, that was all. Other methods of increasing mutation rate were later discovered. These included increases in temperature and the exposure of organisms to certain chemicals.

The study of mutations advanced tremendously as a result, and instead of vague pictures such as the kind drawn in Chapter 5, involving the evolution of lions, tigers and jaguars from ancestral cats, it became possible to study the origin and perpetuation of particular and definite mutations, and apply that knowledge to a better understanding of certain human disorders. In 1946, Muller received the Nobel prize for medicine and physiology for his work with artificially produced mutations.

Of course, the most interesting mutations of all—to us, at least—are those that occur in human beings.

For instance, there is a disease called *hemophilia,* in which human beings have blood that will clot only with difficulty or not at all. This is most dangerous, since a small scratch may result in bleeding to death. A tooth extraction is a major operation, and a bloody nose a disaster.

People with hemophilia have it from birth (which makes it a *congenital disease*) and an investigation of the family tree usually turns up the fact that certain relatives in previous generations have also had it. It is, apparently, the result of a mutation, of a "blood-clotting gene" in which the normal allele produces, through occasional error, an abnormal "hemophilia allele."

Hemophiliacs are under a grave disadvantage and, particularly before the days of modern medical care, there was practically no chance of their living to maturity, let alone having children. It would seem then that the hemophilia allele would die out with those unfortunates who possessed it.

And yet hemophilia does not become extinct. In fact, the percentage of hemophiliacs in the population, though small, remains steady.

For one thing, people can carry the gene without showing it and without harm to themselves. Suppose, for instance, that a person possessed a hemophilia allele in a particular chromosome. There would be another allele of that gene on the chromosome pair, one which in all likelihood would be the normal allele. The normal allele would be dominant and the person with but one dose of the hemophilia allele would be normal. And yet that person might pass on the hemophilia allele to the next generation (each descendant having a 50-50 chance of receiving it). Such a single-dose individual would be a *carrier*.

Now, the interesting thing about the blood-clotting gene is that it occurs on the twenty-third chromosome, the X-chromosome. It is only the woman that has two X-chromosomes, so it is only the woman who can carry a hemophilia allele on one X-chromosome and be saved by a normal allele on the other X-chromosome.

A man has one X-chromosome and one Y-chromosome; the latter carries no genes and is useless to him. If his one X-chromosome carries a hemophilia allele, he has hemophilia. The disease, consequently, shows up almost invariably in males, but it is the female that acts as the carrier.

(This is an example of *sex-linkage*. There are other cases in which a woman's two X-chromosomes give her a margin of safety which man, with his single X-chromosome, lacks. It is to this that some geneticists attribute the fact that more male babies die before birth or are stillborn than female babies; that more male infants die in the first year than female infants; and that even with this weeding out of the weaker males, men generally have a lower life expectancy by at least three years than women do—at least in those sections of the world where the dangers of childbirth have been reduced by modern medical care.)

Hemophilia, then, wipes out the male sufferers of the disease fairly effectively, but nothing happens to the female carriers. The daughters of a female carrier can themselves be carriers, and each carrier can give birth to hemophiliac sons. In the long run, half the sons produced by carriers will be hemophiliac and the other half normal.

Even if it could be supposed that a relentless search were made for probable female carriers and that they were prevented from having children, this would not suffice to wipe out hemophilia. The hemophilia allele is always arising by mutation from normal genes. It is estimated that in every hundred thousand births, one to five new hemophilia alleles are produced, making the infant a hemophiliac, if a boy, and a carrier, if a girl.

(All this points up the difficulties of a eugenics program, since not even the most violent measures will wipe out a disease. Too, there is the question of the normal descendants of hemophilia carriers, who would amount to half in the long run. Many of these might well be very valuable members of society. Do we wish to lose them, along with the hemophiliacs, by the blind application of what we might consider to be eugenic principles? The decisions involved are complicated and difficult.)

Sometimes a detrimental human mutation is kept in being because it is not so detrimental as it seems under certain particular conditions. For instance, examples of such mutations occur in connection with a vital substance in the blood stream, necessary for carrying oxygen from the lungs to the cell. This substance is *hemoglobin*. The normal form of hemoglobin, present in the vast majority of human beings, is *hemoglobin* A.

But there are a few human beings who possess abnormal hemoglobins, a knowledge which came about as the result of a study of a disease first reported in 1910.

In that year, an American physician named James B. Herrick was examining a twenty-year-old West Indian Negro who was suffering from anemia. A drop of the patient's blood under the microscope showed the red corpuscles (which, in normal people, are round and flattish, something like thick coins) to have taken on unusual shapes, many curving like the blade of a sickle. Herrick called them sickle cells, and the condition, *sickle-cell anemia*.

After that, other people, invariably of Negro ancestry, were also found to suffer from such a condition. By 1928, sickle-cell anemia was recognized as an inherited condition (and not caused merely by a faulty diet, for instance, as ordinary anemia is) and a mutated gene was, therefore, at its root.

In 1949, the American chemist Linus Pauling showed that red cells sickled because they contained an abnormal hemoglobin, which he called *hemoglobin* S (the S standing for "sickle cell").

Apparently, the sufferers of sickle-cell anemia are those that have two "S-alleles" of the "hemoglobin-forming gene," one on each of a chromosome pair. The X-chromosome is not involved, so that male and female alike have a margin of safety. If there is but one S-allele, with a normal allele on the pair, some hemoglobin S is formed in the blood, but not enough to cause undue sickling of the red cells. People with such a single dose possess *sickle-cell trait,* which is not fatal or even a great inconvenience, except that two people who both possess the trait may, if they mate, produce children with a double dose, who will have the anemia and probably die young. (In the long run, one-quarter of the children of such unions will have the anemia.)

Again, it is the case of the sufferers dying and the carriers living, but here there is a new factor. Whereas hemophilia seems to strike everywhere and anywhere without discrimination, sickle-cell anemia seems to be confined to people with some Negro ancestry.

About one Negro in eleven in America has the sickle-cell trait, but most American Negroes possess considerable "white" ancestry. A greater concentration of the trait is to be found in certain areas of West Africa where the Negro ancestry is undiluted. There, despite a continuous drizzle toward extinction by the premature death of those with a double dose, the S-allele is maintained at a fairly high level. The level is too high to be accounted for merely by fresh mutations. If mutation were that common, the S-allele would show in other parts of the world as well.

The answer to the puzzle may involve malaria. Of all diseases, malaria is responsible for more human deaths year in and year out in the world generally than is any other disease. Any factor which would make a particular human being more resistant to malarial infection would have a powerful effect in keeping him alive longer than his neighbors. The more common malaria is, in the region where he lives, the greater the effect. Those regions where sickle-cell trait is common are also regions where malaria is endemic, and it would seem that Negroes with sickle-cell trait are more resistant to malaria than those without. (This is reasonable, incidentally, since the protozoan parasite which causes malaria multiplies within the red blood corpuscle, and hemoglobin S may well be less suited to its needs than hemoglobin A.)

There is thus a premium on the possession of a single S-allele. The carrier not only lives a normal life; he lives a better than normal life, at least in regions where malaria is endemic. He has more strength and a longer mating life, so that he leaves more children (which include some anemics who die, but also some carriers who live longer). The advantages of the single dose of the allele tend to increase its numbers, and the disadvantages of the double dose to decrease them. The two tendencies balance to keep the percentage occurrence of the allele at an equilibrium.

Of course, there is this objection. There are regions on the earth where malaria is endemic but where hemoglobin S has not developed to any noticeable extent. Undoubtedly, the mutation occurs there (and for that matter, even in parts of the world where malaria is no threat), but for some reason there is no premium on the possession of a single S-allele, so that the mutation is maintained at an insignificant and unnoticeable level and rarely, if ever, occurs in double dose.

But why should there not be a premium if malaria is endemic? Because to blame it all on malaria is to oversimplify matters. Malaria is one item which happens to be the deciding one in West Africa; it may not be the deciding one elsewhere.

For instance, in Thailand and Indonesia there is a moderately high percentage of occurrence of another kind of abnormal hemoglobin, called *hemoglobin E*. A double dose of this causes serious trouble, but a single dose apparently makes it possible for a human being to get along better when there is but a marginal quantity of iron in the diet. In southeast Asia, the dietary lack of iron apparently imposes the key environmental pressure in this respect, rather than the malaria threat, and favors hemoglobin E over hemoglobin S.

At least a dozen other abnormal hemoglobins have been discovered in the 1950's and very little is as yet known about any of them. The study of human mutations is as yet in its earliest stages.

What causes mutations outside the laboratory? What causes them even when scientists are not tampering with the cell-division mechanism by means of X-rays and chemicals?

Well, even when the scientists' tamperings are eliminated, we are still subjected to mutation-producing conditions. There are the penetrating radiation and speeding particles produced by the violent breakdown of traces of radioactive substances in the soil. There is the penetrating cosmic radiation that bombards earth from outer space. Both of these factors are stronger than X-rays. Even the ultraviolet radiation of the sun, which is weaker than X-rays, but is always with us in quantity, contributes its bit.

Then there are factors we can only guess at. There are chemicals that produce mutations (i.e., *mutagens*), and some have wondered whether exposure to natural occurrences of such might not be responsible for some mutations.

Certainly, an increase in the concentration of any of these factors will increase the mutation rate. Our technical civilization has been increasing the extent to which all of us are more and more exposed to radiation, for instance. For one thing, in the last fifty years there has been an increasing use of X-rays in industry and in medicine. The chance of picking up stray X-rays has increased, and almost all of us (at least in the more advanced parts of the world) are deliberately subjected to X-rays now and then, if only in search of possible tooth cavities or lung lesions.

Then again, the explosion of nuclear bombs in increasing quantities, ever since 1945, has introduced quantities of radioactive atoms into the atmosphere and soil and increased the general intensity of radiation due to radioactivity. At least one type of radioactive atom, *strontium-90,* when absorbed into the body, will settle in the bones and remain there more or less permanently. Before 1945, strontium-90 was nonexistent on earth; today every creature with a bony skeleton or a limy shell contains small quantities of it.

This exposure to increased radiation cannot help but increase the mutation rate. This is a bothersome fact, because most mutations are for the worse. In the long run, to be sure, mutations make the course of evolution move onward and upward, but the percentage of nonmutated or "normal" individuals must remain high to keep a species going more or less normally, while the unwanted mutations are weeded out and the occasional beneficial one is awaited. If the percentage of mutations rises, there is the increasing possibility of the general weakening of a species (or of all species?) past the point where it can be saved by the few beneficial mutations.

Muller, the man who introduced X-rays as a hastener of mutations, is among those who are quite alarmed over this possibility. He not only would like to see an end to nuclear bomb tests (as indeed many would) but would even cut the use of X-rays, themselves, to a minimum and have all people keep thorough records of the occasions on which they have been exposed so that too high a level might never be attained.

The increase in radiation is also important in connection with a type of mutation that does not involve parents and offspring on the organism level. Within a particular organism, there is a parent-offspring relation among the cells. Particular cells are constantly giving rise to daughter cells. In doing so, there is the formation of new genes, and there are imperfections in those formations, too. Mutations that occur from cell to cell are called *somatic* mutations ("somatic" comes from a Greek word meaning "body").

Somatic mutations, which do not involve the gametes, are not passed on to the next generation of organisms. This, however, does not mean that they are unimportant. There is the possibility, for instance, that cancer is caused by a somatic mutation; that somewhere in the body at some time during life, a mutated cell arises through a defect in the formation of a gene and that this mutated cell is cancerous. It passes its defect on to its cellular descendants, and the result is well known.

About the chief reason for supposing that cancer may be the result of a somatic mutation is that those factors which increase the mutation rate also increase the incidence of cancer. It has been known almost since X-rays were discovered in 1895, for instance, that they can cause cancer. (A number of the early investigators of X-rays had to die tragically of cancer to make that fact known.)

The radiations of radioactive substances, even more

penetrating than X-rays, also cause cancer. Both Madame Marie Curie and her daughter Irène Joliot-Curie died of leukemia (a form of cancer), very likely brought on by the continual exposure to the radiations with which they worked. Even ultraviolet light has been found to increase the incidence of skin cancer.

Many of the chemicals that increase the mutation rate also increase the incidence of cancer. Chemicals that increase the incidence of cancer (*carcinogens*) have been found in coal tar, and there have been some who have speculated that modern technology has increased the chemical hazards in connection with cancer, as well as the radiation hazards. Incomplete burning of coal, oil, or tobacco, for instance, could conceivably give rise to carcinogens which might then be breathed in by us. Compounds have recently been located in tobacco smoke that, under some conditions, have been shown to be carcinogenic for certain species of animals. (Presumably, they would be carcinogenic for human beings, too, but there is no direct experimental proof, since experiments to produce cancer artificially in human beings by the use of suspected carcinogens obviously cannot be conducted.) In any case, the possible connection between cigarette smoking and the rising incidence of lung cancer is being most strenuously debated now.

This is an example of the practical results, of such great importance to each of us, that stem from such "ivory-tower" observations as those made by Darwin on finches, by Mendel on pea plants and by De Vries on primroses.

Thus, by the efforts of cytologists and geneticists, the mechanics of evolution were worked out on a cellular level. The questions Darwin could not answer were answered, up to a point.

But only up to a point. It is the nature of science that answers automatically pose new and more subtle questions, and the answers offered by the chromosome theory of inheritance are no exceptions.

For instance, how do genes serve to bring about a particular physical characteristic, anyhow? Just how are new genes formed so exactly like the old, and why and how are they not always formed so exactly?

To pass from Part One to Part Two of this book, we passed from the study of the organism, which is visible to the eye, down to the study of the cell, which is invisible to the eye, but visible to the microscope. Now, to answer the new questions, we pass to Part Three of the book and a new refinement. It is now a matter of considering the molecules within the cell, objects which are not even visible to the microscope. Having gone from the naturalist to the cytologist, we now pass on to the biochemist.

PART THREE

* * * *

11

Building Blocks
in Common

Modern biochemistry began at just about the time that modern geology did, and the two sciences followed oddly similar paths during their first decades.

The nineteenth century began with a great scientist enunciating a wrong doctrine in geology—Cuvier and his theory of catastrophes. A second great scientist performed precisely the same disservice for biochemistry. It came about in the following fashion.

Toward the end of the 1700's, chemists were beginning to recognize two broad classes of substances. One class consisted of those minerals that were found in the soil and ocean, together with the simple gases of the atmosphere. These substances could withstand rough treatment, such as strong heat, without change in their essential nature. Moreover, the existence of these substances seemed to be independent of the existence of living organisms.

On the other hand, there was another class consisting of substances found only in living things or in the dead remains of once-living things. This second class consisted of relatively delicate substances. Under the influence of heat, these chemicals would smoke, char, burn, even explode.

Examples of the first class are salt, water, iron, air, rock; of the second, sugar, alcohol, gasoline, olive oil, rubber, hair.

In 1807 a Swedish chemist, Jöns Jakob Berzelius, suggested that the two classes be named *inorganic* and *organic,* the inorganic being those substances that occurred in nature independently of life, and the organic being those substances produced only by living things.

Berzelius then lent the great weight of his authority (for he was the most renowned chemist of his day) to the belief that the gulf between these two classes was deep and, in part, unbridgeable.

Chemists had learned how to convert one inorganic substance into another in many ways. For instance, they knew how to dissolve zinc in acid to form hydrogen gas. They had also learned to convert one organic substance into another as when sugar was so treated as to ferment and become alcohol. They could even convert an organic substance into an inorganic one, as when alcohol (organic) was burnt to carbon dioxide and water (both considered inorganic).

However, and this was the crucial point, no chemist had yet succeeded in crossing the gulf between the classes in the opposite direction. No one had, by purely chemical means and without the intervention of a living organism, converted an inorganic substance into an organic one. In the opinion of Berzelius, that had not been done because it could not be done.

To produce an organic substance from an inorganic one was, he thought, entirely the province of living organisms; it required the presence of a mysterious "vital force," which the chemist would forever be incapable of duplicating in his test tubes.

For a couple of decades, the theory of the "vital force" held sway, as did the theory of "catastrophes." Then, at almost the same time, both fell. But whereas Cuvier's "catastrophism" fell under the onslaughts of a man of opposing views who popularized an opposing theory already established, Berzelius' "vitalism" was destroyed by one of the chemist's own pupils and by accident.

That pupil was the German chemist Friedrich Wöhler. In 1827, while gently heating a compound called *ammonium cyanate,* which was considered an inorganic chemical, he found that it was converted into another chemical which Wöhler recognized as urea.

This dumfounded Wöhler, for urea was definitely an organic substance. It had been discovered a hundred years before in urine and, in fact, it was the chief solid substance left behind when urine was evaporated to dryness. It was a waste product formed by living organisms and, according to Berzelius' theory, could never be formed from inorganic substances by the ordinary methods of the chemical laboratory.

Yet Wöhler had done just that, and very simply and easily, too. He repeated the experiment a number of times, making certain nothing had gone wrong, that it was indeed ammonium cyanate he began with and urea he ended with. Finally, in 1828, he published his results and there was a sensation. As so often happens in science, it needed only one break in the dike to start a flood. At once, other chemists began to synthesize other organic substances out of inorganic ones.

Berzelius himself was forced to change his mind, and soon chemists were quite satisfied that organic chemicals, although more complex, on the whole, than inorganic chemicals and far more difficult to handle and understand, nevertheless followed the same rules as inorganic chemicals. No one has doubted it since.

"Vitalism" and "catastrophism" fell victim to the same philosophy. The death of "catastrophism" implied that the geologic changes of the past were shown to follow laws similar to those governing the geologic changes of the present. The death of "vitalism" implied that the chemistry of organic compounds and, eventually, of living tissue itself was found to follow the fundamental laws that governed "ordinary" chemicals. It was a double victory for the principles of uniformitarianism.

To be sure, we still divide chemistry into inorganic and organic, but as a matter of convenience only. Organic chemistry is now defined as the chemistry of those compounds containing carbon atoms in their molecules, whether those molecules had ever been formed by any living thing or not. In fact, hundreds of thousands of such molecules have been synthesized in laboratories, some of which do indeed duplicate the compounds of nature, but most of which are completely new and which are substances that, as far as we know, never existed in nature outside the test tubes of the organic chemist.

The narrower concept of organic chemistry—the study of those substances found primarily in living tissue, together with the changes they undergo there—is what is now known as *biochemistry,* the prefix "bio-" coming from the Greek word for "life."

Biochemistry as an empirical study antedated its existence as a science by thousands of years, of course. Even in prehistoric times, people were interested in the properties of food, since it was important to learn how to store it without spoilage, how best to prepare it for eating and so on. And it was impossible to deal with food without becoming acquainted with the different broad classes of substances that comprised it.

For instance, in making bread one dealt with *starch,* a dry, white, tasteless substance insoluble in water. Fruit juice and honey are sweet to the taste and contain *sugar* which, when isolated, is also white but is soluble in water.

The chemists of modern times were not, however, content merely to observe the outer properties of substances. They learned to break them down into simpler and simpler substances and called the simplest (which could be broken down no further), *elements.* Furthermore, it was not enough to identify those elements present in a given substance; each was weighed, and their relative proportions were established. (The chemist who first emphasized the importance of gaining chemical knowledge by accurate weighing and measuring, rather than by simply making note of general properties, was the French chemist Antoine Laurent Lavoisier. He established the idea in the 1780's and it was for this, more than anything else, that Lavoisier is considered the undisputed "father of modern chemistry.")

About 1812, the French chemist Joseph Louis Gay-Lussac followed the principles of Lavoisier and analyzed such substances as sugar and starch to determine the proportions of their elementary content. He found that both sugar and starch contained exactly three elements: carbon, hydrogen and oxygen. In both, the proportions were about the same: roughly forty-five per cent carbon, six per cent hydrogen and forty-nine per cent oxygen, by weight.

The relative percentages, by weight, of hydrogen and oxygen in these substances was just about one to eight, which is the same as the relative percentage of those two elements in water. It seemed, therefore, that an organic compound like sugar or starch might be composed of carbon and water and the same *carbohydrate* ("watered carbon" in Greek) arose. The name is inaccurate, since the structure of the compounds is not as simple as "watered carbon," but it has stuck, anyway.

Gay-Lussac found wood to have an elementary composition similar to that of both starch and sugar and so it, too, is largely carbohydrate in nature. The chief substance in wood is now called *cellulose,* because fibers of it are found mainly in between plant cells, forming stiff and rigid "cell walls."

But carbohydrates are not the only class of substances to be found in food. Another group of substances, perhaps even better known, are the fats and oils. These have a greasy feel, leave translucent marks on paper, are liquid or semisolid, are usually yellowish in color, burn more easily than starches or sugars, and are insoluble in water.

Chemical analysis in the early nineteenth century showed that, like carbohydrates, they were composed of three elements: carbon, hydrogen and oxygen. However, the proportions were different. In fats and oils, the proportions were roughly: seventy-seven per cent carbon, twelve per cent hydrogen and eleven per cent oxygen by weight. Fats and oils are much richer in carbon and much poorer in oxygen than carbohydrates are and no further reason is needed (though many others do exist) for considering two separate classes of compounds to exist here. (In modern times, fats and oils, plus certain related substances, are called *lipids,* from a Greek word for "fat.")

But not all substances in food contain simple carbon, hydrogen and oxygen. An example is a solid substance that can be obtained from egg white. This substance is soluble, but not sweet, and even gentle heating causes it to become insoluble and seems to change its properties radically.

The Latin word for "egg white" was "albumen" (from another Latin word meaning simply "white"), so any substance that behaved like egg white was said to be *albuminous.* Material derived from milk, from blood, and from meat generally was found to be albuminous.

In 1811, the French chemist Claude Louis de Berthollet was able to break down albuminous substance in such a way as to liberate a gas which he recognized as ammonia.

Ammonia is known to contain the element nitrogen. Albuminous substances, therefore, differ from carbohydrates and lipids in possessing a fourth element, nitrogen, in addition to the usual three.

In 1838, a Dutch chemist, Gerard Johann Mulder, went further and analyzed various albuminous substances carefully. He found, in addition to carbon, hydrogen, oxygen and nitrogen, still a fifth element, sulfur. Since the albuminous substances were clearly more complicated than the other types of substances in food, Mulder believed the albuminous class to be more important, as well, and probably the basis of living tissue and of life. He named the class of substances *proteins,* from the Greek word for "first" and, all in all, it has turned out to be a pretty good name—at least until very recently.

At the beginning of Chapter 6, it was said that at the time of Darwin's *Origin of Species* scientists were convinced that "chemically, life was a unit" and that there was, therefore, no chemical reason to doubt that natural selection could convert one species into another. It is now possible to go into more detail.

By the time of Darwin it was quite plain that the organic matter in protoplasm fell, for the most part, into one of the three classes: carbohydrates, lipids and proteins. It did not matter whether the protoplasm was that of an oak tree, a bacterium, an oyster, a snake or a man. Carbohydrate, lipid and protein; that, for the most part, was it.

To be sure, that fact and nothing more might seem to be insufficient grounds for making life a unit or for saying that all protoplasm of all species is essentially alike. Suppose, for instance, we consider the carbohydrates. The starch obtained from potatoes is not precisely like the starch obtained from rice, nor is either identical with the starch from wheat or from bananas. All these kinds of starch are different from the several distinct varieties of sugar, while cellulose is completely different from either starch or sugar. Further, the plant kingdom is rich in cellulose and in a certain sugar called sucrose. The animal kingdom contains neither. On the other hand, another sugar, called lactose, is found only in mammalian milk. It occurs nowhere else in the animal kingdom and certainly nowhere in the plant kingdom.

Is it fair, then, to make much of the fact that all protoplasm contains something so heterogeneous in nature and occurrence as what we choose to call "carbohydrate"?

Actually there is more to "carbohydrate" than just the presence of carbon, hydrogen and oxygen in certain proportions. For instance, in 1812, a German chemist named Gottlieb Sigismund Kirchhoff found that if he heated starch with weak acid, it dissolved. From the solution, he could separate a solid that was no longer starch, but something he recognized as identical to the sugar that was found in grape juice. As it turned out, it didn't matter which variety of the common starches one started with. They all broke down to grape sugar under the action of the acid. (The modern name for grape sugar is *glucose,* from the Greek word for "sweet.")

Then, in 1819, a French chemist, H. Braconnot, found that treating cellulose with acid broke it down to a sugar also—and the same sugar, glucose!

Nor was this coincidence to be confined to the plant kingdom. In 1856, the French physiologist Claude Bernard discovered that the liver of well-fed animals contained respectable quantities of a starchlike substance. It was tasteless in its original form but, on treatment with acid, grew sweet, so he called the liver starch *glycogen* (from Greek words meaning "producer of sweetness"). Why did it become sweet? Because acid converted it to sugar—again the same sugar, glucose.

In fact, glucose itself occurs as such not only in fruit juices such as that of the grape, but (as was first discovered in 1844) also in human blood; and, as we now know, in blood of any variety. It is consequently called "blood sugar" as well as "grape sugar."

In short, the carbohydrates that seem so different are not so different after all. Most of the carbohydrate in the world is built up of a single building block, glucose. The various starches differ only in the number of glucose units built into a single chain; in whether the chains of glucose units are simply straight lines or are branched; and in whether the branches are many or few, long or short. Glycogen, an animal product, differs from starch, a plant product, only in that the branches in the glycogen molecule are unusually many and long and intricately subbranched. Moreover, glucose units can be hooked together in one of two ways. One way results in the various starches and glycogen, the other in cellulose.

Of course, there are some carbohydrates that involve sugars other than glucose. For instance, cane sugar or *sucrose* (which was mentioned before as a purely plant product) has molecules that are made up of one glucose unit combined with a unit of another sugar called *fructose*. In *lactose* (the sugar mentioned as occurring in milk) a glucose unit is combined with a unit of another sugar called *galactose*.

Fructose and galactose, however (together with other simple sugars that occur here and there in living tissue), are very similar in properties to glucose. All these sugars are white, crystalline solids, soluble in water and more or less sweet to the taste. All behave similarly in the presence of certain chemicals; all form yellow insoluble compounds with one chemical, or red soluble compounds with another, and so on.

In other words, the widely dissimilar carbohydrates are made up of a very few, very similar building blocks (of which one predominates), and this is true for all species without known exception.

As for lipids, the case is much the same. In 1811, the French chemist Michel Eugène Chevreul found that lipids broke down in the presence of an acid, forming new, somewhat simpler compounds which had the chemical properties of weak acids. These new compounds he therefore called *fatty acids*.

He distinguished two, which were quite similar in most ways. One is now called *stearic acid* (from a Latin word meaning "solid" because it is found in solid fats particularly) and one *oleic acid* (from a Greek word for "olive oil" because olive oil is particularly rich in it). Other fatty acids were also discovered in the years following. All were quite similar, and the two already mentioned, plus, *palmitic acid* and *linoleic acid*, proved the most common.

All fats and oils, from all species, break down to liberate, for the most part, the same few fatty acids. And in all cases, without exception, these fatty acids are held together by being joined with a substance called *glycerol*.

It seemed fair enough to assume, in Darwin's time, that not only was all protoplasm built up out of the same three classes of substances, but that these classes were in turn built up out of a relatively few building blocks that all species held in common. The completed organisms might be as infinitely various as the completed musical compositions that have been and can be written; but, like the latter, the infinite variety is built upon the arrangement and rearrangement of a relatively small number of notes.

This point of view explained how it was that one form of life could live on another. Potato starch, olive oil, and beef protein are eaten by man and out of them are produced human glycogen, human fat, human protein. What is the mysterious alchemy that brings about the conversion? Nothing more than this: the process of digestion breaks down the foreign food substances to the building blocks that man holds in common with all creatures. It is the building blocks that are absorbed, the simple sugars, the fatty acids and so on; and it is the building blocks that are then put back together again in a fashion that suits our own requirements.

Chemically, all life is one.

Actually, though, I have not mentioned the building blocks of proteins and, in Darwin's time, information about them was not really satisfactory. If at the time of *The Origin of Species* there was anyone who still nourished a faint hope that the chemical differences between the cells of various species were so great as to make the thought of evolution impossible, that last, faint hope had to rest on proteins.

Proteins were more complicated than either carbohydrates or lipids in structure and behavior and were more important to the life processes. As long ago as 1816, it had been shown by the French physiologist François Magendie that dogs could not survive on diets of carbohydrates and lipids alone, but that proteins had to be added. (On the other hand, as was later discovered, animals could live on diets that were almost entirely protein.) It might, therefore, be maintained that carbohydrates and lipids are part of the dead, raw material used by the body, just as water and salt are; and that it was protein that was actually the "life-stuff." If carbohydrates and lipids were built up out of the same simple building blocks in all species, that was of no more importance than that all life depended on water. The question was: how alike were the proteins of the various species?

The first successful attempt to find a building block in proteins was by Braconnot (who had broken down cellulose to glucose in 1819). The next year, 1820, he tried to repeat his success on gelatin, a protein that was obtained by boiling animal gristle. He heated the gelatin with dilute acid, as he had done in the case of cellulose. Eventually, he obtained out of the mixture some white, sweet-tasting crystals, which he naturally took to be sugar, but which was definitely not glucose. He called it "sugar of gelatin." Later that year, he isolated another type of white crystal, which was tasteless and this he called *leucine* (from a Greek word for "white"). Leucine, he found, had nitrogen as part of its structure, which at once made it altogether different from sugars and fatty acids. It was a new kind of building block.

In 1838 it was discovered that Braconnot's "sugar of gelatin" also contained nitrogen and was therefore not a sugar for all its sweetness. It was renamed *glycine* (from the Greek word for "sweet").

In 1846 the German chemist Justus von Liebig heated a milk-curd cheese with an alkaline reagent (one opposite in properties to acids) and out of the resultant material obtained white crystals that were neither glycine nor leucine. He called the substance *tyrosine* (from the Greek word for "cheese").

By Darwin's time, then, there were three entries among the candidates for building blocks of proteins. All contained nitrogen and had certain properties in common but it had to be admitted they did not form the neat, tight group formed by the simple sugars and by the fatty acids. Glycine was sweet, but leucine and tyrosine were not; glycine was quite soluble in water, leucine only slightly soluble, tyrosine practically insoluble. Still, proteins were beginning to yield.

Other crystalline substances were obtained from proteins as the years went by. By the time Sutton presented his chromosome theory of heredity, well over a dozen had been isolated. In some ways they were still a heterogeneous group, but the key point was this: every protein, no matter what its source and no matter what its properties, broke up to give these same building blocks. Glycine, leucine, tyrosine, and all the rest could be obtained out of human protein, whale protein, bat protein, trout protein, snail protein, dandelion protein, bacterial protein. The tyrosine from one was identical with the tyrosine of all the rest, and so on for the remaining building blocks, too.

Furthermore, by Sutton's time a much more subtle way of characterizing substances was found than that of merely listing their outward properties or the proportion by weight of the elements they contained. This new way of looking at substances showed that the building blocks of proteins were by no means as heterogeneous as they seemed.

12

The Shape of
the Unseen

In 1803 the British chemist John Dalton tried to find an explanation for the discoveries that had been made in the previous century in connection with the behavior of gases and simple compounds. He suggested that all matter was made up of tiny, sub-submicroscopic particles called *atoms* (from a Greek word meaning "indivisible," because he supposed there was nothing that could conceivably be smaller). These atoms formed into various groups, now called *molecules* (from a Latin word meaning "a little mass"), and aggregates of these molecules made up visible matter.

Elements, such as carbon, hydrogen, oxygen, iron, gold, phosphorus and so on, were made up of only one kind of atom. *Compounds* were substances with molecules that contained more than one kind of atom. Thus, water had a molecule containing hydrogen and oxygen atoms, while the sugar molecule contained carbon, hydrogen and oxygen atoms.

Dalton, and other chemists, too, used symbols for various elements, since the necessary frequent mention of them was wearying. As each chemist used his own symbols however, this introduced confusion which would have converted all chemistry into an intellectual anarchy, if not stopped.

The man who came to the rescue was Berzelius (whose championing of "vitalism" must not be allowed to obscure the many great contributions he made to chemistry). He suggested, very simply, that elements be symbolized by the initial letters of their names and that when two elements had the same initial letter, a second letter from the body of their names be used. This has been done ever since, and the official symbol (internationally used) for carbon is C, for hydrogen, H, for oxygen, O, for nitrogen, N, for sulfur, S, and so on.

The evidence yielded by several converging chemical investigations showed, for instance, that the water molecule is made up of two hydrogen atoms and one oxygen atom. Its molecule is, therefore, symbolized as H_2O. (Earlier it was said the proportions of hydrogen and oxygen in water were one to eight, but those were proportions by weight; a single oxygen atom is sixteen times as heavy as a single hydrogen atom, or eight times as heavy as two hydrogen atoms.)

Similarly, the molecules of ammonia are composed of one nitrogen atom and three hydrogen atoms (NH_3); of sulfuric acid, two hydrogen atoms, a sulfur atom and four oxygen atoms (H_2SO_4). Such symbols represent *formulas* of the compounds named.

Such formulas, which simply list the number of atoms of each element present in a compound, are quite adequate for the simple compounds that are generally met with in inorganic chemistry. The molecules of organic chemicals are much more complicated, and here unlooked-for trouble arose. For instance, the formula of glucose, the chief building block of the carbohydrates, is made up of six carbon atoms, twelve hydrogen atoms and six oxygen atoms. It can be written $C_6H_{12}O_6$. So far, so good.

Yet fructose and galactose, two other sugars which serve as building blocks, have precisely the same number of the same kind of atoms in their molecules. They, too, have formulas which can be writen $C_6H_{12}O_6$. And yet glucose, fructose and galactose are three different substances with different properties.

This situation, involving identical formulas and differing properties, bothered the early organic chemists, for the situation did not, to their knowledge, arise in inorganic chemistry. Such compounds were called *isomers* (from Greek words meaning "equal parts"), since each isomer contained equal parts of the various elements composing it. This name was first suggested by Berzelius.

The solution of the isomer problem came when chemists realized that it was not merely the number and kinds of atoms in a molecule that counted, but also their arrangement. (The simplest analogy is that of our numbering system, wherein 951 is quite different from 519 or 159, though all are composed of the same three digits.) In the simple inorganic molecules, there was generally only one arrangement possible, so a formula expressing mere numbers and kinds of atoms was enough to distinguish one substance from another. Not so in the more complex organics.

The first who realized this clearly was the German chemist Friedrich August Kekule. In 1858, he began to draw lines between atoms (represented by their symbols) to show their relationship within the molecule. Chemical knowledge already gained showed that the behavior of the molecules could best be explained if the carbon atom was always allowed to form four connections (*bonds*) with neighboring atoms; if three bonds were allowed for each nitrogen atom, two for each oxygen atom, and one for each hydrogen atom. In this way, a *structural formula* could be prepared.

Kekule's system did not mean that structural formulas could at once be prepared for all known organic substances. It took long experimentation and clever deductive work to decide just which structural formulas to use for which isomers; to decide, in other words, on the exact shape of molecules that were far too small to be seen by any known instrument.

It was not until 1891, for instance, that the German chemist Emil Fischer, as a result of detailed and brilliant research (for which he received the Nobel prize in chemistry in 1902) worked out the structural formulas of glucose, fructose and galactose. They look like the following diagram:

GLUCOSE   FRUCTOSE   GALACTOSE

Notice that each formula has six carbon atoms, twelve hydrogen atoms and six oxygen atoms. Notice, too, that the differences in arrangement are minor. Still, out of such minor differences, large variations in properties can result, and upon that, the structure of life can be built.

The structural formula for glycine (which proved to be the simplest of all the building blocks of proteins) is as follows:

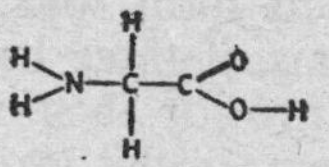

GLYCINE

The group of three atoms on the left, a nitrogen and two hydrogens, is called an *amine group,* because the structure is similar to that of the molecule of the gas ammonia. The group of four atoms on the right, a carbon, two oxygens and a hydrogen, usually lends acid properties to any compound in which it occurs. A compound which possesses both these groups in the molecule, the two groups being separated by a single carbon atom belonging to neither group, as in the case of glycine, is called an *alpha-amino acid.*

As it turned out, all the various building blocks of the proteins, so heterogeneous in some of their properties, turned out to be similar in chemical structure to this extent: they were all alpha-amino acids. The structural formulas of leucine and tyrosine, for instance, are:

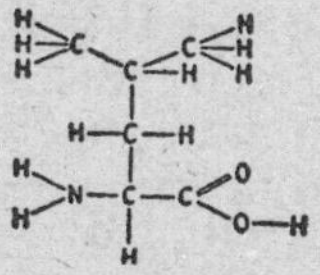

LEUCINE

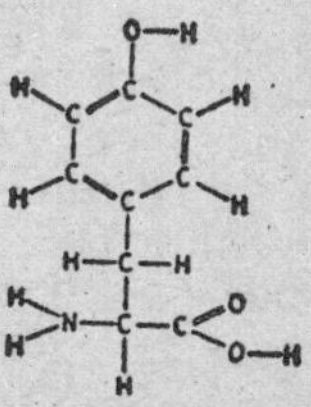

TYROSINE

The difference between these and between either and glycine is that different groups of atoms are, in each case, attached to the carbon atom between the amine group and the acid group. These differing atom combinations are termed *side-chains,* and the different amino acids that make up proteins differ only in the nature of their side-chains.

Emil Fischer (the chemist who had worked out the structure of the simple sugars) also tackled the amino acids. By 1918, he had definitely shown how the amino acids were combined to form protein molecules. The acid group of one combined with the amine group of its neighbor; the acid group of that neighbor combined with the amine group of a third; the acid group of that third combined with the amine group of a fourth, and so on indefinitely. This method of combination held universally among the proteins of all species.

Furthermore, there is a subtlety about amino acids that has not yet been mentioned. It is possible to arrange the atoms of amino acids (with the exception in the case of glycine, the simplest) in two different, but equivalent, ways that are mirror images of each other. This can be shown clearly if the arrangement is made three-dimensionally, using small balls as atomic models. Restricting ourselves to the two-dimensionality of paper, the two ways are conventionally written as follows (with the letter "R" representing the side-chain):

D-AMINO ACID   L-AMINO ACID

The one on the right, again by convention, is called an L-amino acid, the one on the left, a D-amino acid (the letters stand for "levo" and "dextro" from Latin words meaning "left" and "right").

When the chemist tries to manufacture an amino acid in the test tube (which he can do), he always gets a mixture of equal quantities of each form. In the body, however, only one form is synthesized, as was shown in 1924 by the German chemist Karl Freudenberg.

But which form? Both forms are equally stable and have an equal probability of existence. Are there some proteins with one form and some with another? Or do some species use one form while others use another?

No, there is uniformity. The human body contains proteins built up out of L-amino acids only. This is true of all other forms of life, vertebrate and invertebrate, animal and plant, multicellular and unicellular. So uniform is the chemistry of life, that the D-amino acids are often spoken of as "unnatural amino acids," even though they do exist in nature, in small quantities, in a few rare proteins, mostly of bacterial and fungal origin.

There is no known reason why all proteins might not have been built up out of D-amino acids. It is almost as though life began once (let us say as a single cell) and happened to settle on the L-amino acid as a building block, and all life thereafter has been struck with the choice through descent from that original cell.

An example of chemical uniformity that does not involve proteins can also be easily found. By 1911, for instance, various lines of investigation had convinced biochemists that there were hitherto-unknown substances which were necessary, in minute traces, to the functioning of living tissue.

Casimir Funk, a Polish-born American physiologist, suggested that these trace compounds be called "vitamines," the prefix "vit-" coming from the Latin word for "life," and the suffix being an indication that these compounds (as was then thought) contained the amine group. As it turned out, most of the compounds did not contain the amine group, but the name had become accepted and all that could be done was to drop the "e" and change the name to *vitamin*.

Since 1911, well over a dozen vitamins have been discovered, their structural formulas worked out, and their functioning in the chemical machinery of the cell in some cases elucidated. A group of these vitamins, all water-soluble and all containing one or more nitrogen atoms in their molecules, are lumped under the general name of *B-vitamin complex*. Each of the nearly dozen vitamins in the complex occurs, as far as is known, in all cells from human to bacterial; and in all cells, each performs the same function.

In the case of the vitamins, too, all cells seem to share a common inheritance. Chemically, life is indeed a unity.

Yet though life is a unity, it does not represent total

uniformity. There are significant chemical differences that have developed from species to species and these form a kind of *biochemical evolution.*

For instance, all cells (except for those of a few bacteria which do not utilize oxygen) contain certain proteins called *cytochromes* in small quantities. These proteins are colored and their name comes from Greek words meaning "cell color."

Cytochromes are made up of amino acids as are all proteins, but in addition they contain units which are not amino acid in nature. (Proteins built up of amino acids only are *simple proteins;* those which contain non-amino acid material joined with the amino acids are *conjugate proteins* (the word "conjugate" comes from Latin words meaning "joined with"). The non-amino acid portion is a *prosthetic group* (the word "prosthetic" comes from Greek words meaning "placed on to").

The prosthetic group in the case of the cytochromes is a molecule consisting of atoms put together in a rather complicated set of rings called a *porphyrin nucleus* (from a Greek word for "purple," because that is the color of many, but not all, compounds containing such a ring system). When forming part of the cytochromes, the porphyrin nucleus contains an atom of iron attached to certain of its atoms, while groups of carbon, hydrogen and sometimes oxygen atoms are attached at various positions on the ring system. The whole compound is called *heme,* from a Greek word for "blood."

To change the subject for a moment, the key chemical of plant cells is *chlorophyll.* This is a green compound (in fact, the name is from Greek words meaning "green leaf") which enables the plant to form carbohydrates, lipids and proteins from carbon dioxide, water and minerals by using the energy of sunlight. Animals cannot do this, because they lack chlorophyll. All green plants possess chlorophyll; all animals lack chlorophyll. Here is an example of a supremely important and apparently arbitrary chemical distinction that divides all life into two groups.

But what is the structure of chlorophyll? It consists of a porphyrin nucleus in which some of the atom groups attached to the ring have been changed somewhat as compared with heme, and in which a magnesium atom replaces the iron atom. In short, chlorophyll is not something that is completely and amazingly different. It is merely a kind of "mutated heme." Some time in the far distant past, we can imagine, a cell in its manufacture of heme slipped up and manufactured chlorophyll instead, turned it to good use, and from that random occurrence, the plant kingdom developed. There may even have been a series of "pre-chlorophyll" stages in which successively more efficient molecules made use of the energy of sunlight, cells with the more efficient molecules replacing those with the less efficient by the usual forces of natural selection, until modern chlorophyll was developed.

We animals also displayed inventiveness in the same respect. The use of blood (which developed only ages after simple, multicellular life forms had come into being) made it possible to carry oxygen from the outer world to the cells inside the body with new and greater efficiency. For such oxygen transport, a new protein was developed which we call *hemoglobin*. Hemoglobin uses precisely the same prosthetic group, heme (which is named from the Greek word for "blood," by the way), that the cytochromes use. An old compound was thus adapted to a new purpose.

The same line of reasoning which makes it logical to suppose that lions and tigers evolved from a common ancestral species makes it also logical to suppose that the red of blood and the green of leaves evolved from a common ancestral chemical.

Then, too, a chemical distinction can be made between the various classes of the vertebrates. This involves the manner in which organisms get rid of the nitrogen atoms that arise when "used" protein breaks down, as it eventually must, and is replaced by "fresh" protein.

The simplest procedure, apparently, is to combine a nitrogen atom with three hydrogen atoms to form ammonia. This small molecule easily passes through cell membranes into the blood and then out through either gills or kidneys into the ocean beyond.

I say "ocean" because only water-dwelling creatures can make use of this method of getting rid of nitrogenous wastes. Ammonia happens to be extremely poisonous, and if its concentration exceeded one part in a million, the organism would lapse into a coma and die. By dumping the ammonia into the ocean, it is immediately diluted to negligible concentrations. Nor does it accumulate with the eons, because there are microscopic life forms in the ocean that can convert ammonia back into protein eventually, and this protein, directly or indirectly, serves as a food supply for the very organisms that formed the ammonia in the first place.

(This is part of the *nitrogen cycle*. All life is kept going by such cycles, which prevent any chemicals from being truly and permanently used up. In the long run, only solar energy is used up, and there will be ample supplies of that for billions of years to come.)

However, this trick of ammonia elimination fails for creatures living on land. The water supply is not virtually infinite, as it is in the oceans, but is severely limited and must be conserved. To carry off ammonia as quickly as it is formed, without allowing its concentration to rise, would require such quantities of urination as to quickly dehydrate and kill a land organism. For such an organism to survive, it must develop a form of nitrogenous waste which is less toxic than ammonia is and which can, therefore, be allowed to accumulate to higher concentrations and be eliminated by a smaller quantity of urine.

The necessary compound is *urea*, which is formed of the fragments of two ammonia molecules and a carbon dioxide molecule, and ends with a structural formula as follows:

$$\begin{array}{c} \text{H} \\ \text{H} \end{array} \!\!\! N - \overset{\displaystyle \overset{O}{\|}}{C} - N \!\!\! \begin{array}{c} \text{H} \\ \text{H} \end{array}$$

UREA

(It was urea, you may remember, that Wöhler formed from an inorganic compound, thus destroying "vitalism.")

Blood can carry urea up to one part per thousand without disaster, so that urea can accumulate to a thousand times the concentration of ammonia and will require only a thousandth the quantity of water for its elimination. Even that quantity of water can be cut down by allowing urea to accumulate in urine in amounts higher than is possible in blood. In this fashion, it becomes practicable to be a land animal.

Fish eliminate nitrogenous wastes as ammonia; so do tadpoles. However, when a tadpole matures to a frog, it begins to eliminate nitrogenous wastes as urea.

We can see the anatomical changes that take place when a tadpole turns into a frog; the vanishing of the tail and appearance of the legs, the conversion of gills to lungs. We can imagine these to recapitulate the changes that took place over eons as crossopterygian fish evolved into primitive amphibians. However, we cannot see the internal conversion of chemical machinery that converts ammonia elimination to urea elimination, and yet this takes place and is as essential to land life as is the conversion of gills to lungs. This chemical change, too, must be a recapitulation of an equivalent change that took place in the evolution of crossopterygian to amphibian.

Another change was necessary before the great step from amphibian to reptile could be taken. Reptiles do not lay their eggs in water but on land. The water supply within an egg on dry land is even more severely limited than the water supply within a free-living organism on dry land. The organism can at least replenish its water supply by drinking or otherwise absorbing water from outside, but the embryo within the egg cannot.

Yet the embryo must eliminate nitrogenous wastes also, and here even urea would be insufficient. Urea is less toxic than ammonia but the body cannot stand indefinite concentration. What happens in reptilian eggs then, is that four ammonia fragments are combined with three carbon dioxide fragments and two carbon atoms besides, to form a molecule of *uric acid*. Uric acid is a very insoluble substance so that it does not accumulate, to speak of, in the watery contents of the egg, no matter how much of it is formed. It is deposited in small, solid granules in odd corners within the egg, where it does not interfere with the chemical machinery.

Furthermore, in adult life, reptiles continue eliminating nitrogenous wastes as uric acid, so that they have no urine at all, in the liquid sense. Instead, the uric acid is eliminated as a semisolid mass through the same body opening that serves for the elimination of feces. This single body opening is called the *cloaca* (the Latin word for "sewer").

Birds, which developed from reptiles and lay the same kind of eggs, retain the uric-acid mechanism without which egg-laying on land would have proved impossible. They also have the cloaca.

The Prototheria (the primitive egg-laying mammals) also retain both the uric-acid mechanism and the cloaca.

The Eutheria, or placental mammals, which develop the young within the mother's body, do not involve the developing embryo with the extreme shortage of water found in eggs. The entire water supply of the mother is at the embryo's service, and that water supply can be renewed by drinking. Mammalian embryos, therefore, do not need to accumulate uric acid (which is a nuisance to be avoided if possible) and can make do with urea. That is dumped into the mother's blood stream and out of the mother's kidney, leaving the embryo free and clear.

Adult mammals retain the urea system of eliminating nitrogenous wastes and fair quantities of liquid urine are required for the purpose. Mammals lack the cloaca, therefore, but have two separated openings: an anus for solid wastes which comprise the indigestible residue of food, and a urethral opening for the liquid urine.

(As to invertebrates, the water-living forms eliminate ammonia; the land-living forms, such as insects, eliminate uric acid.)

Biochemical evolution is by no means as well worked out as anatomical and physiological evolution. Extinct forms have left plentiful evidence of their anatomy and, indirectly, of their physiology, but no evidence at all concerning their biochemistry.

Nevertheless, all the deductions concerning biochemical evolution which have so far been worked out lead to the conclusions already arrived at independently by those biologists who studied the visible fossil remains of organisms. A nice example, this, of the thesis that there are many roads to truth.

13]

The Surface
Influence

But if the chemistry of life is so uniform from top to bottom, how is it that, in actual fact, a man is so different from a bacterium? They share the same building blocks, the same vitamins, the same chemical abilities.

The answer lies in the fact that, in the last two chapters, there has been deliberate concentration on the building blocks, rather than on the final structures. One might as well ask why Naples differs so from New York in appearance when the structures of both cities are built so largely of stone and brick.

Different proteins are all built of the same amino acids (of which some nineteen are now known to occur generally in the various proteins), connected in the fashion first elucidated by Fischer. However, the amino acids can be arranged in any order, and every different order results in a different protein with different properties.

Each amino acid in the string of amino acids (called a *peptide chain*) making up a protein can be any of the nineteen, without restriction. A peptide chain of, say, five amino acids can, therefore, have any of the nineteen in first place, any of the nineteen in second, and so on. The total number of different peptide chains that can result is $19 \times 19 \times 19 \times 19 \times 19$, or 2,476,099. Two and a half million alternatives, two and a half million different peptide chains with different properties—and only five places are involved.

What if there are 600 amino acids in a protein molecule, as there are, indeed, in a protein with average-sized molecules, such as hemoglobin? The total number would then be so large that to express it we would need a one followed by 639 zeros. Yet even that does not exhaust the possibilities. There are important proteins with thousands and tens of thousands of amino acids in the peptide chain. And there are ways in which proteins can vary in structure and properties even without affecting the actual order of the amino acids. (For instance, the same peptide chain can be wound or coiled in different fashions or linked to other such chains in various ways.)

To all intents and purposes, then, we are safe in saying that the possible number of different protein molecules that can exist is virtually infinite. If every protein molecule ever formed on earth or ever likely to be formed were different, the total number would still form only a complete insignificant fraction of all the different proteins that could theoretically exist.

You might wonder why proteins differ in properties when the amino acid order is changed.

Well, each amino acid has a different side-chain, and when amino acids are combined in the Fischer manner, amine group to acid group, the side-chains are left free and stick out to form the surface of the protein molecule, so to speak.

Now each side-chain has different properties. Some contain few atoms and are small, some contain many and are large. Some carry no electric charge, some a negative electric charge, some a positive. Some are capable of forming certain connecting links with other groups, some are not.

Each different arrangement of amino acids, or each different type of coiling of the same arrangement, results in a molecule with a different surface pattern; one that differs in electric charge pattern, in the manner in which the mechanical projection of groupings is shaped, in the positions where other groups may, or may not, hook on.

These variations in surface can seriously and profoundly affect the organism. Why and how that should be requires some explanation.

In the early days of chemistry, it was noticed that sometimes two substances would react much more quickly in the presence of small quantities of a third substance than in its absence. The third substance, moreover, was not visibly affected by the reaction.

For instance, hydrogen and oxygen will combine with each other so rapidly, when heated, that a mixture of the two gases will explode violently. At room temperature, however, a mixture will stand quietly for indefinite periods without showing any signs of reaction. However, if a bit of finely divided platinum powder is added to the mixture, there will be an explosion even at room temperature. The German chemist Johann Wolfgang Döbereiner discovered this in 1822 and even invented an automatic lighter in which a jet of hydrogen could be directed onto a surface containing powdered platinum, so that it would catch fire spontaneously. (It was a very impressive lighter, but not practical.)

Other such phenomena were discovered, and finally Berzelius, in 1836, discussed all such cases and suggested a name. He proposed calling a substance which, in small quantities, affected the speed of a reaction without itself being affected, a *catalyst,* and the process, *catalysis.* The word comes from Greek words meaning "breaking down," because a catalyst influences the breaking down of other substances.

There is a great temptation to consider catalysis a very mysterious phenomenon, involving some almost supernatural influence of the catalyst upon other substances. Actually, nothing supernatural is involved. A catalyst merely offers a surface upon which a reaction can take place. Such surface influences arise commonly in the ordinary affairs of life and are then taken for granted.

Imagine, for instance, a man with a pencil and paper, and nothing else, standing in the midst of a desert with only soft, shifting sand underfoot. The man wishes to write something upon the paper. He knows how to write, he has something to write with and something to write upon. Nevertheless, he can write only the most fumbling note, one that is very likely to be undecipherable, and he will almost certainly tear the paper in the process.

Now imagine him suddenly endowed with a smooth writing board of polished wood, which will not itself take a pencil mark. In a way, this introduces no change in the situation, since he has no additional knowledge of writing, nothing more to write with, and nothing more on which to write directly.

Yet how different is the situation. Now his message can be written smoothly, clearly and without trouble—all thanks to a writing board, which offers him a surface on which to place the paper so that the writing can take place. The writing board is not itself affected in the process, but remains unchanged. Any number of messages on separate pieces of paper can be written on it, so that one writing board will suffice, given time enough, for a million messages.

The analogy is a close one. A catalyst, too, offers a surface. Finely divided platinum offers a surface on which hydrogen and oxygen molecules can attach themselves and quietly combine at temperatures which, under ordinary circumstances, yield them insufficient energy to combine. When one set of hydrogen and oxygen molecules has combined, it leaves the surface and another set can take its place. Thus, a small quantity of powdered platinum can (and does) suffice for large quantities of hydrogen and oxygen.

As it turns out, catalysis is crucially intertwined with life.

Reactions do take place in living organisms that do not ordinarily take place in the inanimate world. For instance, food is broken down and digested in the stomach, or putrefied by microorganisms. In the absence of action by life, it remains virtually unchanged for prolonged periods. What is there in life that does to food what platinum does to hydrogen and oxygen?

As long ago as 1752 the French naturalist René A. F. de Réaumur, placed meat in little metal tubes with open ends capped by wire mesh and allowed birds of prey to swallow them. Eventually, the tubes were regurgitated by the bird. The metal had protected the food from any mechanical grinding in the stomach, so anything that happened to it was the result of the chemical action of stomach fluids. De Réaumur found the meat gone and only a liquid left behind, so the fluids must have dissolved the meat.

The stomach juices, however, were later found to be acid, and in 1824 the English chemist William Prout showed the acid to be *hydrochloric acid*. This is a strong acid which will act upon meat protein in the test tube and gradually liquefy it. That seemed the answer to Dr.

Réaumur's observations and nothing unusual seemed to be involved.

But in 1835, Schwann (the cell theory man) described experiments which showed that hydrochloric acid alone could not explain the manner in which meat was liquefied in the stomach. He maintained that stomach juices, in addition to the acid, contained some unknown substance which hastened the liquefaction; something that belonged to that class of agents which Berzelius, the next year, was to call "catalysts." Schwann called the new substance *pepsin* (from a Greek word meaning "to digest").

This suggestion was at first met with great skepticism, but then saliva was found to break down starch, despite the fact that saliva contained no acid. Furthermore, juice from a gland known as the pancreas was found to break down proteins, starch and fat, although it contained no acid, either.

And in 1857, Louis Pasteur had discovered that the fermentation of fruit juices (producing wine) was caused by the presence of certain varieties of living cells, called yeast. Without the presence of yeast, the proper fermentation would not go on. More and more, it seemed that living tissue could produce certain catalysts, in the presence of which various reactions, characteristic of life, could proceed.

Because the fermentation reaction was the longest known and best studied of these reactions, it became customary to speak of these life catalysts as *ferments*. Two types of ferments were recognized. One existed outside the cell, the digestive catalysts being examples. If pancreatic juice (filtered and perfectly clear of cells) were removed from the body and placed in a test tube, it would still digest various foodstuffs. Its contents were examples of *unorganized ferments*.

Then there was the kind of catalyst that caused the fermentation of sugar to alcohol. This, it was thought, would proceed only in the presence of intact yeast cells. This was a catalyst that was inseparably bound to life itself and would not exist in the absence of life, and was called an *organized ferment*. (This was a kind of revival, in weakened form, of "vitalism.")

In 1876, the German physiologist Willy Kühne suggested that unorganized ferments be called *enzymes* (from Greek words meaning "in yeast"), to reinforce the notion that they resembled substances in yeast, but to restrict the term "ferment" to the catalysts actually within cells.

In 1897, however, the German chemist Eduard Buchner ground up yeast cells with sand until not one intact cell was left. He filtered off the dead, cell-free juice and showed that it would bring about the fermentation of sugar as well as would the original cells. It was at once obvious that there was no real difference between organized and unorganized ferments and that all catalysts formed by living tissue, in or out of the cell, were merely chemicals and had no mysterious connection with any sort of "vital force." From then on, all body catalysts were included under the name of "enzyme." For this service to science, among others, Buchner received the Nobel prize for chemistry in 1907.

But what were enzymes? What was the structure of their molecules?

The trouble was that enzymes, although essential to the working of living tissue, occurred in such small quantities that it was difficult to isolate them in quantities large enough to study. (Any other substances needed by the body in only small quantities, such as the vitamins and certain minerals, are, as is now known, involved in enzyme action, and are therefore vital to life, although necessary only in traces.)

There was evidence indicating that enzymes were protein in nature, but it was all indirect and many biochemists refused to be convinced. Then, in 1926, the American biochemist James B. Sumner obtained small crystals of some substance from a solution of jack bean flour, and these crystals proved to be a pure enzyme named *urease*. It was a catalyst that hastened the breakdown of urea to carbon dioxide and ammonia. (Again, as in Wöhler's day, urea was involved in a major scientific advance.) When Sumner subjected these crystals of urease to various tests, there could be no doubt that they were protein.

The example was followed. In 1930, another American biochemist, John H. Northrop, crystallized pepsin. He and his research colleagues followed this by the crystallization of still other enzymes.

There are now nearly a hundred enzymes that have been crystallized and all, without exception, have proven to be proteins. It is generally accepted, now, that the thousands of other enzymes that have been studied but not yet crystallized are also proteins. In fact, the word "enzyme" can be defined most simply now as "a catalytic protein," or, if you prefer, "a protein catalyst."

Sumner and Northrop were two of the three men awarded the Nobel prize for chemistry in 1946.

No two enzymes are exactly alike in either function or structure. This is not surprising in view of the virtually infinite possibilities of protein structure variations.

The most natural view of the function of enzymes, once they are known to be proteins and once the infinite variability of the protein molecule is understood, is that they serve as surfaces on which particular reactions may take place. There are many thousands of different reactions going on constantly in all cells, and for each one of those reactions there is a special enzyme, with an amino acid arrangement so designed that its surface is just suitable for the hastening of that particular reaction and few, if any, others.

In general, almost none of the reactions that go on in living cells would proceed at room temperature except with imperceptible speed. The enzymes are, therefore, powerful directors of the chemical machinery. A particular compound might, if left to itself, react very slowly in each of a dozen different ways. Some molecules would follow each of the dozen paths. In the presence of an enzyme, however, which catalyzes only one of those paths, virtually all the molecules would hasten in that catalyzed direction, while virtually none would have a chance to react in the noncatalyzed ways.

The chemistry of a cell is, therefore, the reflection of the type of enzymes it contains, of the quantity of each, and of the position of each within the cell.

For instance, there are small bodies in the cell cytoplasm, called *mitochondria* (from Greek words meaning "cartilage threads," because of their appearance, although they are definitely not composed of cartilage). These mitochondria contain a number of enzymes which serve as catalysts for one step or another of the many reactions that, together, will convert glucose to carbon dioxide and water, liberating energy in the process. It is possible to imagine a glucose molecule as entering the mitochondrion at one end and being passed from enzyme to enzyme, each catalyzing the next reaction until carbon dioxide and water come out the other end, leaving behind, in the mitochondrion, a number of special energy-containing compounds which can be called upon by the cell at any time to liberate energy and thus make life processes possible. (The picture is rather that of an assembly line, in which each enzyme is a worker with a specific function.)

Any interference with any of these enzymes would seriously impair the capacity of the cell to maintain life. In fact, a number of common substances do interfere with one enzyme or another of the mitochondrion and small quantities of these are poisons for that reason. Potassium cyanide is the most familiar example.

It is also possible for an enzyme to be destroyed, without entailing death for the cell or organism, but nevertheless bringing about some radical change.

For instance, most animals have the capacity to form a brownish-black pigment called *melanin*. In human beings, it is melanin that is responsible for brown or black hair, for brown eyes and for swarthiness of skin. Some individuals are rich in melanin, rich enough to have dark-brown skin. Others are poorer in it and have merely olive complexions. Others are poorer still in melanin and are fair-skinned, blue-eyed and blond-haired. Even the fairest normal human being, however, has the capacity to form at least some melanin.

Now melanin is produced from tyrosine (the amino acid first isolated in cheese) as the result of a number of successive chemical reactions, each of which is catalyzed by some appropriate enzyme. One of these enzymes (the one catalyzing the first step in the process, as a matter of fact) is called *tyrosinase*. Occasionally, a human being (or other organism) is born without the ability to form tyrosinase. Without tyrosinase, the entire series of reactions forming melanin comes to a halt.

An individual without tyrosinase, therefore, has white hair and skin, and eyes that are colorless except for the color of blood showing through. Such individuals are *albinos* (from a Latin word for "white") and the change from the normal condition is most striking, considering that it has come about through the loss of but a single one of the many thousands of enzymes present in human beings.

To be born an albino can be the result of a mutation, just as being born a hemophiliac may be. The parents of albinos may be quite normally pigmented individuals. They may even be Negroes.

In view of cases such as this, it was inevitable that sooner or later geneticists would turn their attention to the inheritance of enzymes. Beginning about 1941, the American geneticist George W. Beadle did just this. He worked with a pink bread mold called *Neurospora,* which ordinarily requires nothing more than some sugar and minerals (plus one vitamin) to live on. Naturally, it makes use of all the usual amino acids in its proteins but it manufactures these in all necessary quantities out of the sugar and minerals.

Beadle exposed the *Neurospora* to ultraviolet radiation and to X-rays to encourage mutations and, sure enough, he obtained a spore which would not grow in the sugar-mineral solution. It might, however, grow if he added, say, a particular amino acid to the nutrient mixture. Once the spore started growing, its appearance was no different from that of a normal mold specimen. Nevertheless, it was a mutant, since it lacked some enzyme that served to synthesize the amino acid in normal specimens. Without the ability to synthesize it, the mutant had to have the amino acid supplied it ready-made and would not grow without it.

By this method, Beadle could follow mutations involving *Neurospora* enzymes as well as Muller could follow mutations involving *Drosophila* wing shapes..

Beadle could even gain knowledge about enzymes that was available to biochemists in no other way. For instance, he would try to grow a mutant *Neurospora* spore on various compounds resembling the amino acid it required. If we call these *precursors* (that is, compounds that might be formed by the organism on the way to the formation of the amino acid) A, B, C, and D, it might turn out that the mutant would grow on C and D, but not on A and B. The conclusion would be that the mutant possessed enzymes that would convert C and D to the amino acid but not A and B. If B resembled C closely and if biochemical experience indicated that the type of change involved in going from B to C usually required a single enzyme, then it would be possible to say that the *Neurospora* lacked the enzyme catalyzing the conversion of B to C.

A second mutant, also requiring the same amino acid in the diet, might be able to grow on B as well as on C and D, showing that *Neurospora* could indeed have the B-to-C enzyme. However, this second mutant might not be able to live on A, indicating the loss of an A-to-B enzyme.

By the study of many such mutations, it was possible to work out detailed schemes for the synthesis routes of many amino acids, vitamins and other compounds of biochemical importance. It was also possible to show definitely that the presence or absence of enzymes was a gene-controlled characteristic, following the ordinary laws of genetics.

In fact, geneticists now more or less accept the fact that genes exert their influence through the enzymes they cause to be formed (or fail to cause to be formed). The enzyme pattern of a particular organism gives rise to its various physical characteristics. Sometimes the connection can be traced, as in the case of albinism; much more often, the connection is obscure. But, obscure or not, the connection is there.

But if the enzyme pattern is controlled by the gene pattern, the obvious question is: how?

How does a particular gene supervise the formation of a particular enzyme (or a particular group of enzymes, perhaps) and not any other enzyme out of all the infinite number possible?

To answer that question it is necessary to consider a type of substance quite different from any of those yet discussed.

14

The Living Molecule

In 1869 a German chemist named Friedrich Miescher was working with pus (broken-down white blood cells) and obtained from it a material which was neither carbohydrate, lipid nor protein. It was made up of carbon, hydrogen, oxygen and nitrogen, as proteins were, but in addition contained phosphorus. Because white blood cells usually have very prominent nuclei and because Miescher suspected this new substance came from those nuclei, he called it *nuclein*.

Nuclein showed definite acid properties, however, and by 1889, it became customary to speak of it as *nucleic acid* and that has been its name ever since. It was also found that within the cells nucleic acid was associated with protein and the two together formed a substance which was called *nucleoprotein*.

Biochemists were at the time most interested in protein, and investigation of the nucleic acid portion of nucleoprotein molecules proceeded slowly. It was found that nucleic acid broke down on treatment with acid to yield smaller building blocks, just as was true of other large molecules in living tissue. In the case of the nucleic acids, the building blocks were called *nucleotides*.

The nucleotides, themselves, could be broken down further; and each was found to consist, in its turn, of three parts. One part was phosphate (the phosphorus-containing portion), another a sugar, and the third a nitrogen-containing compound of a rather unusual type.

The chief investigator of nucleic acids in the early days was a Russian-born American chemist, Phoebus Aaron Levene. In 1911, he showed that the sugar contained in the nucleotides of one type of nucleic acid was *ribose*. This sugar had been synthesized by Emil Fischer back in 1901 and he had invented the name "ribose" for it, without its having any particular meaning. It was considered a laboratory sugar that did not occur in nature, until Levene showed otherwise. Levene also found a second type of nucleic acid with nucleotides containing a sugar similar to ribose but with one oxygen atom missing. This he called *deoxyribose*. The formulas for the two sugars follow.

RIBOSE   DEOXYRIBOSE

If you were to compare these formulas with that for glucose given earlier in the book, you would see that these differ mainly in that they have one less carbon atom.

In any particular nucleic acid, the nucleotides are always identical with respect to the sugar component; either they all contain ribose, or they all contain deoxyribose. Nucleic acids are divided into two species, so to speak, for that reason; they are called *ribosenucleic acid* and *deoxyribosenucleic acid*. For the sake of convenience, these names are usually abbreviated (like government agencies) and are spoken of as, respectively, *RNA* and *DNA*.

Eventually, it was discovered that RNA occurred chiefly in the cytoplasm of the cell, with only minor quantities present in the nucleus. DNA, however, was present only in the nucleus, and never in the cytoplasm. (In the case of RNA, "nucleic acid" is obviously a misnomer, but the name sticks.)

The third component of the nucleotides, the nitrogen-containing compounds, was found to vary in structure. The atoms composing the molecules of these compounds are arranged in distinctive rings, sometimes in a double, sometimes in a single, ring.

The double-ring variety belongs to that class of compounds known as *purines*. This name was given these compounds by Emil Fischer back in 1881, partly because he was pleased to obtain them in "pure" form, and partly because of their connection with uric acid (uric acid is itself a purine).

The two purine compounds present in nucleic acids are *adenine* (first obtained from glandular tissue, hence its name, which comes from the Greek word for "gland") and *guanine* (so called because it was first obtained from the bird excrement called "guano"). Their structural formulas, showing the double ring of atoms, follow.

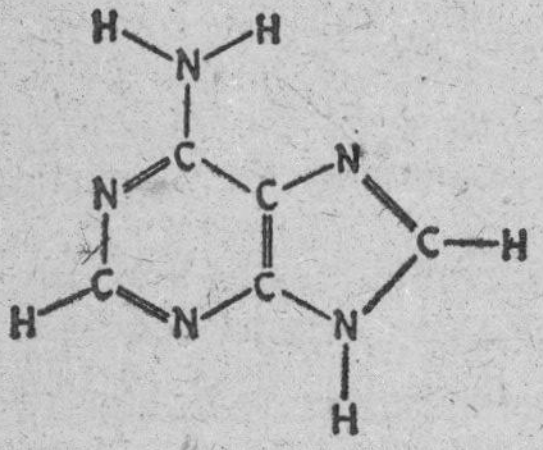

ADENINE 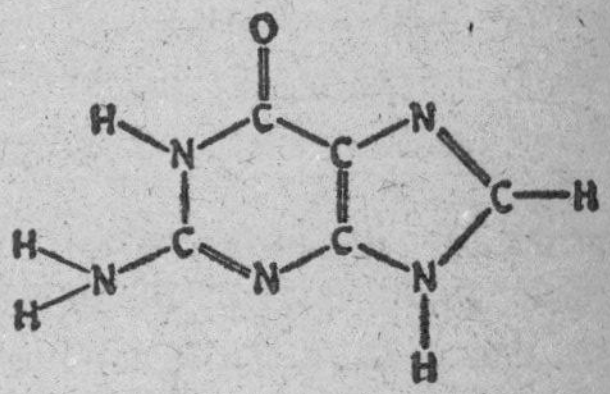   GUANINE

The single-ring compounds are *pyrimidines* (a name with a complicated derivation not worth going into). Molecules of DNA contain two different pyrimidines, *cytosine* (from the Greek word for "cell," since all cells contain DNA) and *thymine* (since the thymus gland is particularly rich in it). Their structural formulas follow.

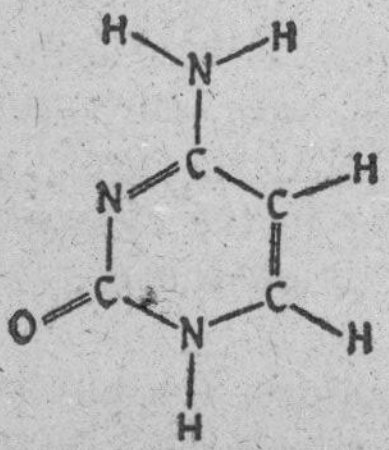

CYTOSINE 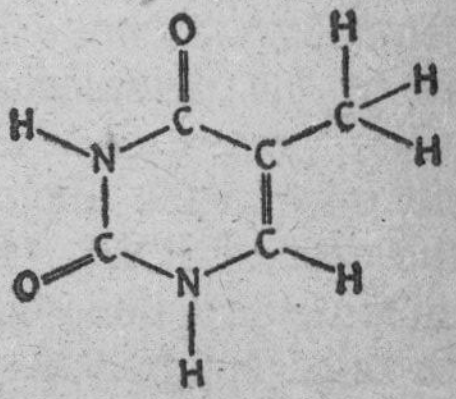   THYMINE

Molecules of RNA do not contain thymine. Instead, they contain a very similar pyrimidine called *uracil* (a name which is a kind of abbreviation of uric acid which it also resembles somewhat in structure). The formula of uracil follows.

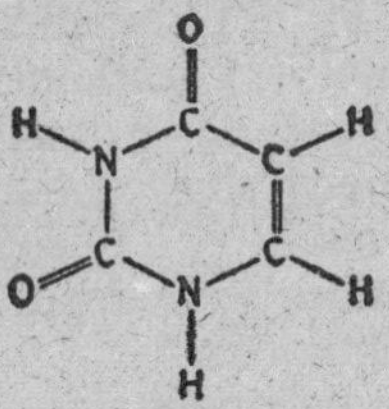

URACIL

The manner in which nucleotides are hooked together to form nucleic acids took far longer to work out than did the similar problem of amino acids hooked together to form proteins. It now appears that in the individual nucleotide the nitrogenous compound (N) is connected to the sugar (S), which is connected to the phosphate (P). The phosphate group of each nucleotide is connected also to the sugar of the neighboring nucleotide, and that *internucleotide link* holds the nucleic acid together. A schematic diagram can be made of what a nucleic acid molecule must look like, as follows.

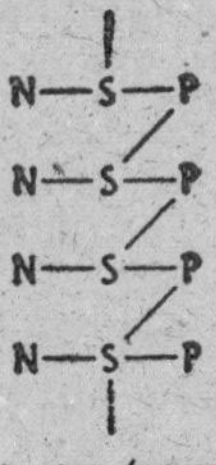

NUCLEIC ACID (SCHEMATIC)

In any given nucleic acid molecule, the phosphate group is the same all down the line of nucleotides. So is the sugar group, being either ribose (in RNA) or deoxyribose (in DNA). The nitrogenous compounds can vary, however, being any one of four, and each of the four occurring at different places down the line of nucleotides.

Levene, in fact, considered the nucleic acid molecule to be made up of just four nucleotides, one containing each of the nitrogenous compounds. This would make the nucleic acid molecule rather smaller than that of a lipid and much smaller than those of starch and proteins. This view was held well into the 1930's.

However, beginning in 1939, studies of nucleic acids extracted from tissue in a very gentle manner (so that the molecules would not break down into fragments in the very process of extraction) showed that more than four nucleotides must be present per nucleic acid molecule. First dozens of nucleotides were reported per molecule, then hundreds, then thousands.

By the 1950's, it was generally accepted (rather to the surprise of most biochemists) that in its natural state within the cell the molecule of nucleic acid was as large as any protein molecule, and larger than most. It could be made up of a thousand or more nucleotide units strung together.

Slowly the unique importance of nucleoproteins began to be realized by biochemists. As stated earlier in the book, staining methods had been developed to color some parts of cells and not others, and the importance of chromosomes was first understood through the use of such stains. But the stains that colored chromosomes also colored nucleoproteins!

Several lines of evidence all began to converge toward the view that the chromosomes were nucleoprotein in nature. About 1936, the matter was virtually settled when a Swedish biochemist, T. Caspersson, devised a method for taking microphotographs of a cell illuminated by ultraviolet light. The purines and pyrimidines in nucleic acids absorbed such light, while most of the other cell constituents did not. Regions containing nucleic acids, therefore, showed up white against a black background. It turned out that both chromosomes in the nucleus and mitochondria in the cytoplasm contained nucleic acids. The nucleic acid in the chromosomes was almost entirely DNA; that in the mitochondria, entirely RNA.

The thought then arose that the genes, which until then had been rather mysterious units whose existence was only deduced from genetic data, might be definite chemical compounds; that they might, in fact, be merely complex nucleoprotein molecules.

A second and completely independent line of investigation also pointed to the nucleoprotein molecule as being chiefly implicated in inheritance.

Louis Pasteur, in 1862, first published his *germ theory of disease,* which stated that infectious diseases were caused by the parasitic activity of microscopic organisms within the human body. The first great advances in the control of infectious diseases came in the decades following the establishment of Pasteur's theory, as physicians learned to isolate the bacteria causing a particular disease and then found some way to fight them, by means of a chemical, vaccine or serum.

Yet bacteria were not always found that could be associated with a particular disease. Pasteur himself studied the disease hydrophobia, and produced a vaccine against it, yet could find no microscopic agent that caused it. Pasteur was too confident of his theory to allow this one fact to overthrow it. He simply pointed out that the causative agent of hydrophobia was probably too small to be seen by a microscope.

Another disease with no visible causative agent was "tobacco-mosaic disease," an infectious condition in which the leaves of tobacco plants grew mottled. In 1889, a Dutch bacteriologist, Martinus Willem Beijerinck, referred to whatever invisible agent or poison carried that disease as a *virus,* a word which in Latin means simply "poison."

In 1892, the Russian botanist D. Ivanovski made a mash of leaves from tobacco plants suffering from the disease and passed the liquid from the mash through filters so fine that even bacteria could not pass through. The bacteria-free liquid that emerged could, however, still pass the disease on to healthy tobacco plants. The agent was, therefore, a *filterable virus.*

In 1916, the American bacteriologist H. A. Allard used a finer filter which would hold back even some particles too small to be seen under the microscope. Liquid passing through such a filter did not cause the disease. The virus, then, was a particle too small to be seen by ordinary microscopes but larger than the protein molecules which could pass through Allard's filter.

Then, in 1935, the American biochemist Wendell M. Stanley separated out pure tobacco-mosaic virus and crystallized it. Once this was done, viruses could be studied as substances that could be weighed and subjected to definite chemical tests. It was found at once that tobacco-mosaic virus was nucleoprotein in nature. It contained RNA.

Later on, when other viruses were crystallized, all were found to be nucleoprotein in nature. Some contained only RNA, some only DNA, some both.

Because of this work, Stanley shared the 1946 Nobel prize for chemistry, along with Sumner and Northrop, the crystallizers of enzymes.

Viruses come in all sizes and complexities, but all have this property in common: they cannot grow and reproduce independently, they can do so only within some living cell. It is as though they are themselves incomplete cells.

It is tempting to suppose that viruses may have grown incomplete as a result of their turning to parasitism. Higher organisms which turn to parasitism specialize by losing structures and organs they no longer need as parasites. Perhaps cells that have turned to parasitism also lose various cellular structures and chemical abilities.

This cellular degeneration can be seen in stages. For instance, there is a group of microorganisms called Rickettsia (named after the American pathologist Howard Taylor Ricketts, who first discovered them) which cause such diseases as typhus fever, psittacosis and Rocky Mountain spotted fever. The Rickettsia are sufficiently large to be seen in the microscope but they are smaller than ordinary cells and are apparently incomplete. At least they can only grow and reproduce within the cells they parasitize.

Below the Rickettsia are the viruses proper, which have abandoned more and more of their cellular properties. There are large viruses, like *vaccinia* (which cause smallpox) and *bacteriophage* (which infest bacteria) which still retain substances in the fashion of possible free-living ancestors. They possess, for instance, certain phosphorus-containing fatlike compounds called *phospholipids,* and even some enzymes and vitamins.

As viruses grow smaller and smaller, however, more

and more of this extraneous material is abandoned until the tiniest viruses (like the tobacco-mosaic virus) are nucleoprotein only. It is as though life had finally gotten down to the bare chromosome; as though the smallest viruses were nothing more than collections of "wild genes" ready to invade cells and impose their own will upon them over and above the supervision of the cell's own genes.

(A variation on this theme is the fact that certain filterable particles were discovered which could pass on some forms of cancer from one organism to another of the same species. The first example was discovered in 1911 by Peyton Rous. Others have been discovered since. These are called *tumor viruses,* and are also nucleoprotein in nature. A tumor virus might be looked upon as a "mutated gene," and a cancerous one, that can be transmitted from cell to cell.)

All of this made it look as though it was nucleoprotein that was the essential of life, and that all else was merely commentary. The other substances in the cell were the machinery, so to speak, with which the nucleoprotein worked; it was the nucleoprotein that did the working. The converging evidence that both gene and virus were nucleoprotein seemed to make that plain.

But why nucleoprotein rather than any other kind of protein? There seemed, for many decades, no particular importance of the nucleic acid prosthetic group over other types of prosthetic groups. Even when, in 1939 and thereafter, it turned out that nucleic acid molecules were very large, biochemists were not unduly impressed. Mere size is not all-important. Starch molecules can be extremely large, but they are made up of only one type of building block, glucose. Their properties lack the flexibility and versatility of the protein molecule, which is built up of nineteen building blocks, the various amino acids. Nucleic acids, built up of four building blocks (two nucleotides containing purines and two containing pyrimidines), might be more versatile than starch, but must be far less so, almost infinitely less so, than proteins. Or so it seemed.

Yet evidence piled upon evidence to show that nucleic acids were important, far more so than they seemed to be.

The first bit of evidence came from the sperm cell. The history of that evidence dates back to the very beginning of nucleic acid chemistry.

It was Miescher himself (the discoverer of nucleic acids) who first isolated nucleic acid in the sperm cells of fish. (This in itself was really the first indication that chromosomes are nucleoprotein in nature, since, as was said earlier in the book, sperm cells are little more than tiny bags of compressed chromosomes.) Along with the nucleic acid, Miescher discovered a protein which seemed to contain many amine groups in the side-chains of its amino acids, so he called the protein *protamine*.

When the sperm cells of other species were studied, it was found that protamines were not universally present. Another type of protein found along with nucleic acid in sperm cells was one called *histone* (from the Greek word for "tissue"). Sperm cells always contained either protamine or histone.

But here arose an odd point. Histones, by and large, are rather simple proteins, considerably simpler than most cell proteins. The molecules are smaller, and certain amino acids predominate, which ordinarily do not. Protamines are simpler still, being, in fact, so simple as scarcely to seem true proteins. A typical protamine molecule might be made up of a string of only about seventy amino acids, as compared with six hundred or so in a protein like hemoglobin, which is itself only of average size. Further, more than fifty of the amino acids in the peptide chain of a protamine molecule might be of a single variety, a type of amino acid called *arginine*.

This seemed difficult to understand. The sperm cell carries all the genes necessary to transmit all the inherited characteristics. It must follow, then, that unusually simple proteins, ridiculously simple proteins, suffice to contain within their structure all the supremely complicated paraphernalia of inheritance.

This difficulty was pressing enough to cause the German biochemist Albrecht Kossel (who in 1910 was awarded the Nobel prize in medicine and physiology) to suggest, in 1897, that protamines were the nucleus about which protein molecules were built. This implied that if a particular protamine was present in a sperm cell, a particular protein in full complexity could be built up out of it once the sperm cell had safely entered the ovum and was surrounded by raw material. This fully complex protein might then carry the information of inheritance.

This suggestion was never fully accepted. Nevertheless, biochemists retained for a long time the notion that (in order to save weight, perhaps, so that the sperm cell might be more maneuverable and be quicker to reach the waiting egg cell) all but the absolutely necessary portion of the protein molecule was discarded in sperm-cell formation. Just enough was carried along, in the form of histone or even merely of protamine, to serve as a foundation upon which the rest of the gene could be built up within the fertilized egg. (The protein that is associated with nucleic acid in the chromosomes of ordinary cells is, as it eventually turned out, a fully complicated one.)

However, once the true size of the nucleic acid was understood, it became plain that, despite any necessity for weight reduction, all sperm cells carried along a set of nucleic acid molecules in full size, as complex as any in ordinary tissue. There was no simplification whatsoever. It was as though sperm cells could afford to skimp on protein, but not on nucleic acid. Or perhaps (and this thought must have started occurring to some biochemists in the early 1940's) it was the nucleic acid portion of the nucleoprotein molecule that carried the genetic information and not the protein portion.

A stronger hint next came in the study of bacteria. A particular species of bacteria (or virus) can exist in several *strains,* that is, in several varieties distinguished from each other by their appearance, virulence, infectivity, or any other measurable characteristic. New strains continually arise through mutations.

In the case of the pneumococcus (a bacterium causing pneumonia), two strains are called "smooth" and "rough" because the first variety possesses a complex carbohydrate capsule as enclosure, which gives the bacteria a smooth appearance under the microscope. The second variety, not possessing this capsule, appears rough.

Experimenters discovered that an extract of the smooth variety could be prepared which, if added to the rough variety, converted the rough variety into a smooth one, indistinguishable from the ordinary smooth variety. It was as though the rough strain lacked a certain gene, which could be mechanically added from the smooth strain that possessed it. Once the gene was added, the rough strain became, and remained, smooth, passing the gene on to its descendant cells until such time as a mutation, involving the loss of the gene, once again gave rise to a rough strain.

In 1944, a group of American biochemists, including O. T. Avery, showed that the chemical in the extract that changed the strains was DNA. It was pure nucleic acid, with no protein whatever in the extract. The DNA behaved like a gene without the help of any protein. DNA was carrying genetic information, all right; it contained within its structure the ability to supervise the formation of some enzyme that made it possible to construct the capsule of the smooth variety.

For the first time, biochemists found themselves forced to consider nucleic acid as material that might possibly be even more important than protein.

Decisive evidence in this same direction came in connection with the study of viruses, which has been proceeding with great vigor in the 1950's.

In 1955, for instance, the biochemist H. Fraenkel-Conrat reported that he had treated tobacco-mosaic virus in such a way that it was separated into a nucleic acid portion and a protein portion, neither of which was too badly damaged. Although the tobacco-mosaic virus had been infective before, neither fraction separately was infective. Neither protein alone nor nucleic acid alone seemed capable of giving the disease to the tobacco plant.

If the two fractions of the original virus were then mixed, a certain amount of infectivity, about one per cent of the original, was restored. Apparently, the nucleic acid rejoins the protein in the mixture but usually in an incorrect and useless manner. One out of a hundred reunions, however, clicks into place correctly and the intact virus molecule is restored.

In one respect, however, things are not quite so simple. Actually some infectivity remained in the nucleic acid portion of the virus molecule. It was very small in amount, and at first it seemed it must be due to the nucleic acid portion being impure. There must be, it seemed, a small quantity of intact virus hanging on.

But no amount of purification of the nucleic acid seemed to remove the small residual infectivity, and doubts grew. Was the nucleic acid infective on its own but just having trouble getting into the cell? The nucleic acid from the virus was injected into cells and, sure enough, there it multiplied.

The situation with respect to nucleic acid and protein within a virus would seem to be analogous to that of a man and an automobile. The man is alive and can travel from city to city by himself, but slowly and with difficulty. With the aid of a nonliving automobile he can do the job easily and quickly. The nonliving automobile cannot do the job by itself at all.

This shows itself (with respect to a virus) even more plainly in the case of bacteriophage, the bacteria-infesting virus. (Bacteriophage was discovered by the Canadian bacteriologist Felix Hubert D'Hérelle in 1915, when he found that certain cell-free liquids apparently dissolved bacterial cultures and wiped them out.)

Under the electron microscope (a device which uses streams of high-speed electrons, rather than light rays, to magnify objects, and which can make visible many objects far too small to be seen under ordinary microscopes) bacteriophage proved to be a comparatively large virus. The most common strains of the virus are shaped like tiny tadpoles, with a polyhedral head and a distinct, stubby tail.

Bacteriophage was studied by X-ray diffraction; that is, by aiming a beam of X-rays at it and observing the manner in which this beam was turned from its path. It was also bombarded with various subatomic particles to see how and in what manner the virus particles were damaged. From such studies (using techniques that were unknown and unheard of a short generation ago) the American bacteriologist Ernest C. Pollard drew a picture of the virus in the mid-1950's that is now generally accepted.

The protein of the bacteriophage forms a hollow shell

on the outside of the virus. Inside the hollow is coiled the nucleic acid. When a bacteriophage molecule encounters a bacterial cell, the end of the bacteriophage tail is attracted strongly to a specific spot on the bacterial cell surface. (Probably the pattern of electric charge on the virus tail-tip just matches, in reverse—for unlike charges attract—the pattern of electric charge on the surface spot.)

In any case, the virus makes contact and sticks. A digestive enzyme at the tip of the bacteriophage tail now catalyzes the dissolution of that portion of the bacterial cell surface with which it is in contact.

That, you see, is the essential service performed by the protein of the virus. It takes the nucleic acid to the cell interior, as the automobile takes the man to the next city. But once man and automobile arrive, it is the man, and not the automobile, who must fulfill the purpose of the trip; this also holds for the nucleic acid protein. After arrival at the cell interior, it is the nucleic acid that takes over.

The nucleic acid moves into the bacterium. Only the nucleic acid moves inside; the protein shell of the virus remains outside!

This startling turn of affairs is a conclusion that is deduced from several converging lines of evidence. For one thing, the protein of the virus can be so treated as to incorporate into its structure atoms of radioactive sulfur, while the nucleic acid of the virus is made to incorporate atoms of radioactive phosphorus. Both types of radioactive atoms can be easily and unmistakably detected and distinguished by modern instruments. After bacteria have been infected with the bacteriophage, the radioactive phosphorus is found within the bacterial cell; it cannot be removed without completely disintegrating the cell. The radioactive sulfur, on the other hand, remains on the outside of the cell. It can be washed off, or even shaken off, while leaving the cells intact and unbroken.

Within the bacterial cell, the nucleic acid of the bacteriophage acts as though it were a foreign gene that has successfully invaded and conquered the cell. It takes over the duties of the cell's own genes. It is the invading nucleic acid that now supervises the chemical machinery of the cell. Under the forced rule of the invader, the cell machinery turns out replicas of the virus nucleic acid, and not replicas of the nucleic acid molecules of the bacterium itself, as should be the customary task of the cell.

Not only that, but the bacteriophage nucleic acid also forces the bacterial cell machinery to form bacteriophage protein as well. When the process is complete and the sucked-dry bacterial cell dissolves into shreds, there are present hundreds upon hundreds of complete bacteriophage molecules, each with its deadly nucleic acid coiled within its protein shell.

There seems little doubt now that the nucleic acid, rather than the nucleoprotein, is the irreducible (as far as we can tell) essential of life. Nucleic acid, in its natural state, is what might be called a living molecule. All else, including all protein, is but the machinery it works with.

But how does nucleic acid work its machinery? How does it manage to enforce a replica of itself upon the raw material of the cell contents? How does it supervise the formation of protein molecules which are so unlike itself?

This is a subject deserving of a new chapter.

15

Passing on
the Information

When Levene first worked out his theory of the structure of nucleic acid (that is, that it was made up of one of each of the four different nucleotides), one of his reasons for doing so was that analysis seemed to show that equal quantities of each nucleotide could be isolated from nucleic acids. In the late 1940's, using *paper chromatography,* a technique unknown in Levene's time, this was proved to be not quite so.

(In paper chromatography a small quantity of a mixture of similar compounds is placed at one end of a sheet of filter paper and is allowed to dry there. An appropriate liquid is allowed to creep up the paper, past the spot and beyond. As the liquid creeps on, it drags the components of the mixture with it, but each component is dragged at its own characteristic rate. After a while, the individual components have separated, like runners in a race who begin abreast but, because of their differing speeds, end in single file. Each of the components [which may be the individual nucleotides of a nucleic acid], may be removed from the paper separately and its quantity determined. This technique, developed in 1944 by a group of British biochemists, including A. J. P. Martin, is now the most important single technique in biochemistry. It is hard to think of any branch of biochemical research that does not use it or how the science could progress further without it.)

Paper chromatography in the hands of such biochemists as Erwin Chargaff showed, by 1949, that the four nucleotides were never present in quite equal proportions in nucleic acids. However, some regularities did show up. The total number of purines seemed always to be roughly equal to the total number of pyrimidines. That meant that adenine plus guanine was equal to thymine plus cytosine in DNA (or to uracil plus cytosine in RNA). Furthermore, the number of adenine nucleotides was roughly equal to the number of thymine (or uracil) nucleotides and the number of guanine nucleotides equal to the number of cytosine nucleotides.

Furthermore, by breaking down nucleic acids carefully and observing the make-up of the fragments, it became clear in the early 1950's that there was no set order or periodicity to the arrangement of the nucleotides. Just as amino acids can (and do) occur in any order in proteins, so nucleotides can (and do) occur in any order in nucleic acids. For the same reason that no two proteins need be alike, so no two nucleic acids need be alike.

In 1953, two biochemists at Cambridge University, F. H. C. Crick and J. D. Watson, using X-ray diffraction data, deduced that molecules of nucleic acids in viruses (and presumably elsewhere) consisted not of one, but of two nucleotide strands. This double strand was arranged in a helix about a common axis; that is in the form of two interlocking, spiral staircases about the same central post. The two strands were so arranged that the purines and pyrimidines of one faced the purines and pyrimidines of the other, each purine (or pyrimidine) being attached to the purine (or pyrimidine) opposite by a type of weak link called a *hydrogen bond*.

The hydrogen bond is only a twentieth as strong as the bonds that usually hold atoms together within a molecule. It is strong enough, even so, to hold the two strands in place. Yet it is also weak enough to break and allow the two chains to separate on occasion, without requiring more energy for the purpose than the cell can easily supply.

The distance between the strands is equal throughout. The gap is too narrow to allow two of the comparatively large purine molecules to face each other and yet too wide for two of the smaller pyrimidines to face each other. The only possibility that fits the facts is that all down the line of thousands of nucleotides, a purine on one strand faces a pyrimidine on the other and vice versa. This would at once account for the observation that the total

number of purines in nucleic acids seems to equal the total number of pyrimidines.

Furthermore, if it is assumed that an adenine on one
strand is invariably faced with a thymine on the other
strand, while a guanine on one strand is invariably faced
with a cytosine on the other strand, that would account
for the additional observation that the number of adenines
in the nucleic acids seem to equal the number of thy-
mines, while the number of guanines seem to equal the
number of cytosines.

None of this interferes with the randomness of the nucleic acid structure. Each strand, taken by itself, can have
any arrangement of nucleotides. It is only with respect
to each other that the strands show anything other than
randomness. The structure of one strand determines the
structure of the other; they fit together like a plug and
a socket or one jigsaw piece and its neighbor.

If we want to simplify the Watson-Crick picture to the
fullest, we can let adenine be represented by A and
thymine (or uracil, in RNA), its invariable partner, by a;
guanine by B and cytosine, its partner, by b. Now the
double-stranded nucleic acid can be schematically represented in the accompanying figure, in which the helical
shape is straightened out for the sake of convenience and
the hydrogen bonds are represented by slanted, rather
than horizontal, lines for reasons that will be clear later.
Notice that the hydrogen bonds invariably connect an A
with an a and a B with a b, but that the order of A, $a B$,
and B down either one of the strands is as random as I
could make it.

The problem of replication of the nucleic acid now
lends itself to a dramatic solution.

Suppose that conditions within the cell are such that
the weak hydrogen bonds are broken and the double
helix of the nucleic acid separates into two single strands.
Suppose, further, that each separate strand is surrounded
by cell fluid which contains (as it does) a plentiful supply
of individual nucleotides, or material out of which nucleotides can be formed on short notice.

These individual nucleotides are always (by blind movement) striking against the single nucleotide strands. If a
thymine nucleotide strikes a section of the strand carrying

an adenine nucleotide, it attaches itself by a hydrogen bond. If it strikes any other section, the hydrogen bond does not form. Similarly, a cytosine nucleotide will attach itself to a guanine nucleotide. The same is true in reverse in both cases.

In other words, *A* will fit itself to *a, a* to *A; B* to *b;* and *b to B.* When all the nucleotides are lined up, each to its natural mate, they are combined into a chain by the action of appropriate enzymes.

In short, strand *x* acts as a mold for the formation of an adjoining strand just like the strand *y* that left it. Similarly, strand *y* acts as a mold for the formation of an adjoining strand *x.* Each is the basis for a new double strand and the result is that two (each exactly like the first) exist afterward where only one double strand existed before.

Clear as this picture is, it is not without its problems. For instance, just how do the two nucleotide strands manage to separate? On a molecular scale, they make up a long and intimate union and it is not easy to see how there could be enough time for the two strands to get completely disentangled before starting to replicate. Furthermore, once loose, the individual strands ought easily to get twisted and fail to behave as proper molds.

A suggested possibility designed to get around this difficulty is that as the chains start separating at one end, the proper nucleotides start hooking on at once. The new joinings proceed down the line as the separation continues, and replication is complete the moment the disentanglement is. It is as though you imagined a slide-fastener opening and, as it opened, a new row of teeth joining to each of the separating halves so that when you completed the opening you found, not one open slide-fastener, but two identical closed ones.

Even this, however, does not answer the question of what it is that spurs the strands of the double helix into a separation in the first place when mitosis is beginning and not at other times. I know of no suggested answer to this question, but then, how dull science would become if all questions had answers already known. Fortunately, on that basis, science will never be dull.

Using the Watson-Crick model of replication, one can see how a nucleotide strand might, on occasion, fail to

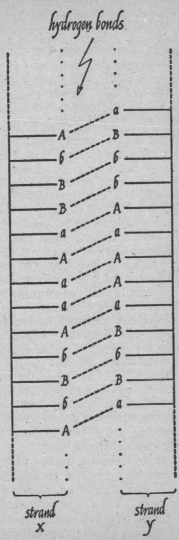

hydrogen bonds

strand x

strand y

THE DOUBLE STRAND OF NUCLEIC ACID (SCHEMATIC)

replicate itself perfectly. For instance, *A*, rather than *b*, might just happen to wedge itself next to *b* and be built into a strand in that place before it could bounce away. The result would be a double strand with an unusual *b*—*A* combination in place of the ancestral *b*—*B*.

At the next replication, the *b* of the first strand would attach itself to a *B* as it should and become a double strand of the ancestral type. However, the interloping *A* of the other strand would attach itself to an *a*, forming an *a*—*A* double strand that would be unlike the ancestral type and which would replicate itself thereafter as an *a*—*A* until such time as another imperfect replication would involve it.

At any given point of the double strand, such imperfect replications would happen rarely. However, there are thousands of nucleotide pairs along the strand, and for an imperfection to take place somewhere among the thousands is perhaps not so rare.

Each imperfect replication would produce a slightly different nucleic acid. The slightly different nucleic acid would produce in its turn a slightly different enzyme which would introduce a slight difference into cell chemistry which, in the long run, might produce some deviation from the normal great enough to be seen by eye. An imperfection during nucleic acid replication would through a chain of circumstances, in other words, result in a mutation.

Another way of looking at it is to suppose that actually any nucleotide could join to any other, but that there is a minimum of energy that the body needs to supply to cause *A* and *a* to join, and *B* and *b* to join. In that case, a new strand in which *a* and *A* join and *b* and *B* join all down the line requires the least energy and is the most probable situation. Imperfectly matched strands require more energy to form, are therefore less probable, but can nevertheless be formed. The more imperfect the matching, the less probable it is, and the more infrequent the occurrence.

If, then, extraneous energy is supplied to the cell so that there is more energy available, imperfect matches (which are energy-consuming) are now more likely to take place. A supply of energy, in the form of elevated temperature, ultraviolet light, X-rays, radioactive particles, and so on, could, therefore, bring about an increase in the number of imperfect replications and consequently, increase the rate of mutations.

Specific enzymes that are involved in the process of nucleic acid replication have been isolated. In 1955, the Spanish-born American biochemist, Severo Ochoa, isolated one (from a bacterium) which was involved in the formation of RNA from nucleotides. In 1956, a former pupil of Ochoa's, Arthur Kornberg, isolated an enzyme (from another bacterium) which could bring about the formation of DNA.

By supplying such an enzyme with a particular nucleotide as raw material, Ochoa found he could form synthetic RNA made up entirely of nucleotides containing uracil groups or adenine groups and so on. He could even build up synthetic nucleic acids containing different nucleotides.

Kornberg could do the same for DNA and went even a step further. He added a bit of natural DNA to act as a "mold" upon which the enzyme might form new DNA and showed that the DNA thus formed was identical with the DNA used as mold. In 1959, Ochoa and Kornberg shared the Nobel Prize for medicine and physiology, in consequence.

The Watson-Crick theory of nucleic acid replication was quickly adopted by most scientists. A few difficulties arose, to be sure, but were easily handled. Thus, some viruses contained DNA molecules made up of a helix with a single strand (first reported by Robert L. Sinsheimer of the California Institute of Technology). How could a single strand replicate itself?

Call it X. It would produce its opposite number, x, which would then produce its opposite number, X. In other words, X would replicate itself in two steps rather than one. That process is less efficient than the one-step replication of the double-strand, which is why it has survived only in certain very small viruses.

The problem that really troubled those interested in the Watson-Crick theory involved the matter of protein production. As an example, virus nucleic acid not only replicated itself, but also produced virus protein, which looked nothing at all like itself. How was that managed?

What's more, the virus nucleic acid did not produce just any protein, but a particular protein only. It produced the specific virus protein itself, which is different from any protein in the cell infected by the virus, or, for that matter, from any protein belonging to any other kind or strain of virus. And, as a related question, how does the nucleic acid in our own genes bring about the formation of a particular enzyme with a particular protein structure, and no other?

Somehow, the nucleic acid molecule must contain within itself all the information required for the construction of a protein molecule.

Consider, for instance, that the strands of a nucleic acid molecule are made up of a series of different varieties of nucleotide arranged in any of an enormous variety of orders. Suppose that each different variety of nucleotide guided the placement of one particular kind of amino acid. The body could then produce a protein with particular amino acids in the same arrangement as the particular nucleotides of the DNA.

Unfortunately, matters can't possibly be as simple as that. There are only four different nucleotides making up nucleic acid molecules and there are twenty different amino acids making up protein molecules, and that could mean trouble.

It is as though you were trying to form a code by letting digits represent letters: 0 standing for A, 1 for B, 2 for C and so on. Therefore, we could write 201 for "cab" and 2064 for "cage" and so on. But we would run out of digits when we reached 9 = J, and we would have no system for representing "rosy" in this code. The code would break down.

The solution is very simple, though. We can use combinations of digits. If we start with 10 = A, 11 = B, 12 = C and so on, we could reach 19 = J and have plenty of combinations left over for punctuation.

In fact, you don't need more than two different symbols to make a very useful code. The Morse Code has only a dot and a dash to work with, and yet proper combinations of these can represent all twenty-six letters, plus the ten digits and the various punctuation marks. And by using twenty-six different letters and no more, we can build up all the hundreds of thousands of words in the English language.

In 1954, the Russian-born American physicist, George Gamow, did suggest that the nucleic acid molecules carried their information in nucleotide-combination. His actual design for those combinations proved wrong, but it set people thinking.

There are four different nucleotides, but if you take them in pairs you have a total of four times four, or sixteen different items. This still isn't enough to take care of nineteen different amino acids, but if you take the nucleotides in triplets, you have a total of four times four times four, or sixty-four different combinations and that is certainly enough.

The total number of pairs and triplets is given in the accompanying diagram (see page 198).

We can imagine, then, that a particular nucleic acid molecule will bring about the production of a particular protein by means of the order of nucleotide-triplets in the chain. If we have (AaB) (abb) (BBb) (ABa), these may stand for four particular amino acids arranged in that order.

This device, whereby a nucleotide triplet stands for an amino acid, is called "the genetic code," and a nucleotide triplet is sometimes called a "codon" for this reason.

You may think that sixty-four codons is too many for nineteen amino acids, but each codon doesn't have to stand for a different amino acid. Two or even three different codons can stand for the same amino acid; thus (ABb), (Abb) and (Aab) may all stand for one particular amino acid. This would be what mathematicians would call a "degenerate code" but it is a perfectly good one just the same.

In fact, it might help to be a little degenerate. If the body made a small mistake, in replicating the nucleotide combinations and produced Abb instead of ABb, the same amino acid could be formed and there would be no damage. This possibility of getting the correct product despite mistakes in the procedure is useful and common. It exists in the English language and our system of writing. If I write "the genntic code," you will have no trouble seeing what I mean despite the misspelling. (In fact, you may not even notice the misspelling at first.)

The Four Nucleotides

A	B	a	b

The 16 Nucleotide Pairs

AA	AB	Aa	Ab
BA	BB	Ba	Bb
aA	aB	aa	ab
bA	bB	ba	bb

The 64 Nucleotide Triplets, or Codons

AAA	AAB	AAa	AAb	ABA	ABB	ABa	ABb
BAA	BAB	BAa	BAb	BBA	BBB	BBa	BBb
aAA	aAB	aAa	aAb	aBA	aBB	aBa	aBb
bAA	bAB	bAa	bAb	bBA	bBB	bBa	bBb
AaA	AaB	Aaa	Aab	AbA	AbB	Aba	Abb
BaA	BaB	Baa	Bab	BbA	BbB	Bba	Bbb
aaA	aaB	aaa	aab	abA	abB	aba	abb
baA	baB	baa	bab	bbA	bbB	bba	bbb

But this raises the question of which particular codon stands for which particular amino acid. We must "break the code" and find out.

The first step in this direction came in 1961 as a result of an experiment by Marshall W. Nirenberg and J. Heinrich Matthaei at the National Institutes of Health.

They began with a synthetic nucleic acid molecule which they had built up out of only a single variety of nucleotide. It could be represented as AAAAAAAAAAAA. . . . Therefore it could be made up of only one kind of codon: (AAA) (AAA) (AAA) (AAA). . . .

They then set up a system whereby proteins could be formed. They added the necessary amino acids and other cell components required for the job. Sure enough, they obtained a protein out of the mixture—and a very peculiar kind of protein, too. It was a protein made up of a single amino acid endlessly repeated. They knew at once that that codon, AAA, stood for that particular amino acid, and the first item in the genetic code "dictionary" had been obtained.

Little by little, in the years that followed, further items in the dictionary were added, until by the middle 1960's, all the codons had been identified as standing for some particular amino acid and no other. What's more, the same dictionary served all species from viruses to man. The genetic code was universal.

Scientists have even worked out the details of how the instructions contained in the genetic code gets from the cell nucleus (where the DNA of the chromosomes is to be found) out to the cytoplasm where the enzymes are manufactured.

A strand of the DNA molecule in a chromosome can form an opposite number out of the nucleotides that go into RNA. This RNA molecule, stamped with the DNA structure and therefore carrying the DNA "instructions" travels into the cytoplasm. It is "messenger-RNA" for it carries the DNA message.

In the cytoplasm, messenger-RNA attaches itself to tiny particles called "ribosomes" where the proteins are formed. There the messenger-RNA acts as a mold on which that formation can take place.

In the cytoplasm, there are also present small molecules of RNA, so small as to be freely soluble in the cell fluid. They were first detected by the American chemist, Mahlon B. Hoagland, and were called "soluble-RNA."

The soluble-RNA is a double-headed molecule. At one end, it has a codon; that is, a nucleotide triplet. Naturally, this codon will fit only onto the opposite version of itself on the messenger-RNA molecule. A soluble-RNA molecule with AAA at one end will only fit on a spot on the messenger-RNA molecule where there is an aaa, and so on.

This means that each variety of soluble-RNA (and there are as many varieties as there are different codons) will fit only into certain places of the messenger-RNA molecule. You will end by having a line of different soluble-RNA molecules in an order dictated by the order of codons in the messenger-RNA (which, in turn, is dictated by the order of codons in the DNA molecules in the chromosome).

The other end of the soluble-RNA molecules can attach itself to an amino acid, but each different soluble-

RNA molecule will attach itself only to a certain amino acid. If the various soluble-RNA molecules are lined up in a certain order, according to the structure of the messenger-RNA, their other ends consist of amino acids lined up in a certain order. The codon information is thus transferred to the amino acids, and soluble-RNA is therefore also called "transfer-RNA." Then, when the amino acids on the other end of the transfer-RNA molecules are knit together, a protein molecule is formed; a *particular* protein molecule with a structure dictated by the structure of the DNA in the chromosomes.*

It is possible that all this theoretical concern with the inside of the cell may end by introducing startling changes in man's way of life.

We are reaching the point where we can perhaps begin to turn from molecular hunting to molecular herding.

The human race did something like that, once, on a large scale. Carnivorous man was first a hunter, foraging for what game he could find and going hungry when he could not find it. At some stage, however, he learned that if he kept certain animals behind fences or under guard and fed them and took care of them, they would breed. Instead of going out to search for animals, his tame herds would produce animals for him. Some of these would serve to keep the herd going, while the surplus would serve as food.

On a molecular scale, however, we are still hunters. If we want the hormone insulin, for instance, we must look for it in its native haunt, the pancreas. The pancreases most available are those of slaughtered cattle and swine. However, each steer and each hog has one, and only one, pancreas so that there is an upper limit to the amount of insulin that can be available in a given time.

If we must have more insulin we are out of luck!

But suppose we "tame" the nucleic acid molecules which, in the appropriate pancreas cells, supervise not merely the production of more nucleic acid like itself, *à la* Kornberg, but the manufacture of insulin molecules as well, what then? What if we put these nucleic acid mole-

---

* If you would like to read about all this in greater detail, you can refer to my book *The Genetic Code,* which is also published by Signet Science Library.

cules in a test tube at the right temperature and in the right surroundings and feed them amino acids (which can be prepared by the ton, if necessary)? There would then be no theoretical reason why we could not have any amount of insulin prepared for us. Or any other protein, following the same principle. We would be herding protein, not hunting it.

This is just at the edge of being something more than speculation and dream. In 1958, scientists at the California Institute of Technology, under the leadership of R. S. Schweet, used certain cytoplasmic particles called *microsomes* as their herd. They derived these from immature red cells, which are in the business of making hemoglobin at a great rate. The nucleic acid which supervises hemoglobin manufacture is in the microsomes. The biochemists added certain energy-containing compounds to the microsomes plus appropriate amino acids and, sure enough, found hemoglobin produced.

There is even the possibility that we may be able to change Nature, or even improve upon it. We improve on Nature with respect to our ordinary herds, breeding special varieties of domestic animals to improve the yield of meat, milk, eggs, wool, and so on. Can something analogous to this be done with proteins?

Well, in 1958, V. G. Allfrey and A. E. Mirsky of the Rockefeller Institute reported that they had isolated cell nuclei and removed the nucleic acid therefrom. At once the ability of the nucleus to manufacture protein came to an end. If the nucleic acid were replaced, the ability to manufacture protein was restored.

What was particularly startling, however, was that if, instead of nucleic acid, a synthetic polymer (resembling the nucleic acid only in having a long-chain molecule with similar distribution of electric charge) were added to the nucleus, protein manufacture was restored.

Could it be that someday, by making appropriate polymers, we can create new types of proteins designed to do special jobs for us?

Perhaps. We can speculate, at any rate, and in modern science even the wildest speculations have a way of sometimes coming true.

Life

PART FOUR

* * * *

16

In the Beginning

From all that has been scaid in the third section of the book, it would certainly seem that the one-celled animal, with which the sketch of evolutionary development began in Chapter 6, is by no means the beginning of life, after all. Primitive though a unicellular creature seems in comparison to a man, or even to an oyster, it must be the end product of a long line of evolution, of which no trace has been left.

Is there enough time for that? Astronomers currently believe that the universe is six billion, possibly even twelve billion, years old. The sun, and the solar system generally, is perhaps five billion years old.

These enormous lengths of time are not, however, fully available for the development of life on earth. Life on this planet could only have developed after the earth's solid crust was laid down and after the oceans were formed.

Yet even that limits us to no mean interval of time. The oldest rocks in the earth's crust (as judged by the slow radioactive decay of the uranium they contain) seem to be about three and a half billion years old. The crust (and presumably the ocean) is, therefore, that old at least.

The earliest fossils we know of are not much more than half a billion years old but, to be sure, the multicellular forms of life then existing were already quite advanced.

Even if we double the time and allow a full billion years as the time during which multicellular life has been in existence, there would still be a gap of two and a half billion years between the time of the forming of earth's crust and its ocean and the development of multicellular life. Two and a half billion years during which cells might slowly evolve from subcellular life and in which those cells might develop and grow complex! Ample time, in all probability.

From a consideration of the chemical make-up of the universe as a whole, and of the solar system in particular, it would seem that the earth's original atmosphere could not have been at all like the atmosphere it now has. The original atmosphere must probably have been composed of compounds rich in hydrogen, since it is estimated that the universe is about ninety per cent hydrogen and the sun is eighty-five per cent hydrogen.

As an example of a hydrogen-rich atmosphere, consider that of the planet Jupiter, which is mostly hydrogen and helium (in the proportions of three to one) with minor quantities of hydrogen-containing gases such as ammonia (NH_3) and methane (CH_4). The atmospheres of the other giant planets beyond Jupiter are similar.

The earth is much smaller than Jupiter, however, and the earth's gravitational field is not strong enough to hold on to the very light molecules of hydrogen and the almost-as-light atoms of helium. However, the field could hold on to methane and ammonia and the earth's original atmosphere may well have contained these, plus sizable quantities of carbon dioxide. There would be no free oxygen in this atmosphere.

Since carbon dioxide and ammonia are both quite soluble in water, the earth's original ocean must have been loaded with those two compounds. Furthermore, both air and water would have been exposed to a much harsher sunlight than we are exposed of today.

The sun emits a rich variety of ultraviolet rays, but these react with the oxygen in the upper reaches of our present atmosphere, forming a particularly energetic variety of oxygen, which is called *ozone*. Almost all the ultraviolet radiation of the sun is absorbed in the process, and goes into the maintenance of the ozone layer (or *ozono-*

sphere fifteen miles above the surface of the earth. Very little of the ultraviolet radiation actually penetrates down to the surface, which is a good thing, for the sun's full supply would kill us. (Even the feeble quantity of the less energetic variety that does reach us can result in painful burns to the fair-skinned and unwary.)

In the earth's primordial atmosphere, however, where no free oxygen would have existed, there would have been no ozone formation. All the ultraviolet rays of the sun would reach the surface, or almost all. The energetic ultraviolet light, bombarding the ocean and dense surface atmosphere, would have supplied the energy necessary to convert the simple molecules of water, carbon dioxide, methane and ammonia into more complicated molecules, and still more complicated ones. (Earth, in primordial days, possessed more radioactivity than it does now, and radioactive radiations may have helped, too.)

In 1952, an American chemist, S. L. Miller, circulated a mixture of water, ammonia, methane, and hydrogen past an electric discharge for a week, trying to duplicate primordial conditions (with the electric discharge representing the energy supply of ultraviolet light). At the end of the week, he found organic compounds in his solution that had not been there to begin with. Even some of the simpler amino acids were present—and he had been working only a week.

Without being able to be certain, of course (and perhaps we never will be), we can speculate as to the possible course of events in the primordial ocean.

Under the drive of energy, ultraviolet or radioactive, the primordial ocean would have slowly filled with more and more complex compounds: amino acids, sugars, porphyrins, nucleotides. These would be built up further so that amino acids might combine into proteins, and nucleotides into nucleic acids.

This could continue at random for perhaps a billion years or more, until a time came when a double-stranded nucleic acid molecule was put together which was complex enough to have the capacity of consummating replication in the manner described in the previous chapter. To have this happen on the basis of random chance seems to be asking a lot, but then a billion years is a long time.

And if this indeed happened (and surely something like it must have), then at least once in the history of our planet, there did, after all, take place a case of spontaneous generation. It was a stupendous event, too, the most stupendous in the history of our planet, for by it all of life may have been formed in one split-second of random synthesis.

Once such a nucleic acid molecule was formed, the equivalent of free-living genes (or tiny viruses) were present in the ocean, and they multiplied at the expense of the organic compounds that had been built up all about them by the action of the sun. These original viruses were not parasites, for there was nothing for them to be parasitic upon.

Eventually, an equilibrium was reached, for as the organic compounds which served as food were incorporated into the virus molecules, food concentration became progressively thinner, and it became ever more difficult for the original viruses to multiply. In the long run, the viruses could multiply no faster than the sun could build up a food supply for them; finally their numbers would become stabilized, and the ocean would contain a thin scum of life.

In the process of replication, there would be numerous imperfections, so that eventually there would be many strains of viruses, each with somewhat different capacities. Natural selection would play its role and those viruses which could compete most successfully with their fellows for the thinned food supply would replicate most frequently. Their strains would become predominant, and in this fashion there would be a slow evolution of viruses.

For instance, some viruses might stick together after replication, forming a string of individual nucleoprotein molecules. The individual molecules might specialize as a result of imperfect replications now and again, and pass on their specializations to descendants when the entire group replicated at once. Those groups in which the specializations best fitted, making the most efficient whole, multiplied at the expense of the other, less efficient nucleoprotein groups, and also, of course, at the expense of the individual nucleoprotein molecules. In this way, the equivalent of free-living chromosomes (or large viruses) then swarmed in the ocean.

The pressure of a depleting food supply must have placed a high premium of survival on any virus that managed to store food more efficiently than its neighbors. Some strains may have somehow developed a membrane about themselves through which small molecules like sugar and amino acid could pass, but not large molecules like starch and proteins. Such viruses could absorb small molecules and build them up into large molecules which would be trapped within the membrane. They would have succeeded in accumulating a food supply and preserving it for their own exclusive use. They would survive at the expense of the naked, improvident viruses, and thus the ocean would become filled with very primitive cells.

These cells must have put the precellular organisms out of business. It is possible that some subcells survived by giving up the fight for an independent competition for food (so to speak) and turning to parasitism as an out. They let the cells collect the food, then invaded the cells and lived on them. Or, as may be more likely, none of the subcells survived, but some of the less efficient cells found the going too rough in competition with the more efficient ones and themselves turned to parasitism, gradually losing their cellular specializations and forming the whole gamut of parasitic viruses of today, from the Rickettsia on down. In either case, the hypothetical free-living viruses of the primordial ocean were wiped out.

Certain cells then developed chlorophyll, which enabled them to manufacture starch and proteins from the simple molecules all about them (from water, carbon dioxide and some minerals), using sunlight as the source of energy. These were the first plant cells.

As pointed out in Chapter 6, plant cells no longer depended on the slow formation of food by ultraviolet radiation, as had the preplants. Instead, plant cells manufactured their own food and could multiply to many times the numbers that had previously been possible. Those cells that did not develop chlorophyll could indirectly benefit also, for instead of scouring the ocean for the thinning supply of organic material, they could let the chlorophyll-containing cells manufacture food and then eat those cells, food and all. The development of chlorophyll, in short, made it possible for the ocean to grow thick with life.

Chlorophyll and the *photosynthesis* ("putting together by light") that it made possible inevitably altered the nature of the atmosphere. When carbon dioxide and water are combined to form starch by the action of chlorophyll, there is oxygen left over which is discharged into the atmosphere as oxygen gas. The carbon dioxide is slowly used up and oxygen takes its place. The growing amount of oxygen combines with the ammonia in the atmosphere and oceans. It combines with the hydrogen atoms in the ammonia molecule particularly, forming water and leaving the nitrogen atoms of the ammonia molecule to combine in pairs to form gaseous nitrogen and remain behind in the atmosphere. The oxygen also combines with any methane present to form carbon dioxide and water, the carbon dioxide being broken down further to oxygen.

The end result is the formation of our present atmosphere of oxygen and nitrogen.

An atmosphere containing free oxygen must have completely revolutionized life, since oxygen is a powerful chemical that requires careful handling. Life forms had to develop cytochromes, for instance, to handle it. The only life forms that exist without cytochromes today are certain *anaerobic bacteria,* which live without oxygen and to which, indeed, oxygen is poisonous (one of the best known of these is the germ causing tetanus, or "lockjaw"). Perhaps the anaerobes are the last remnants of ultra-conservative life which succeeded in filling an environmental niche that still resembles what all of the earth must have been like in the days before chlorophyll.

In Chapter 12, chlorophyll was described as possibly having arisen from the "mutated heme" of a cytochrome molecule. This is one possible way of looking at the matter since almost all creatures, plant and animal alike, possess cytochromes, while only plants possess chlorophyll. It is the cytochrome that would thus seem more fundamental and the earlier formed.

However, if the oxygen atmosphere is indeed the result of photosynthesis, then perhaps matters are reversed. The heme of cytochromes may be a "mutated chlorophyll." In that case, animals must have developed the heme of cytochromes independently from some molecule (now lost) that was ancestral to both chlorophyll and heme, or else animals must be descended from some primitive plant forms.

In fact, in the last couple of years, the latter suggestion has indeed been made. The most primitive plant forms now existing are the blue-green algae. They are so primitive that they lack a clearly defined nucleus or chromosomes. The only other life forms that are simpler in structure are the bacteria and the viruses. (Sometimes the blue-green algae, bacteria and viruses are put into a separate kingdom on the basis of their primitive structure.)

It has been suggested that all creatures whose cells (whether one or many) possess well-developed nuclei are evolved from the blue-green algae (or from their ancestors, rather). These creatures with well-developed nuclear cells include other forms of algae, plus multicellular plants, plus all animals, whether one-celled or many-celled.

The original animal cells may be looked upon as offshoots of the early blue-green algae. The algae had developed first chlorophyll (which produced the oxygen), then cytochromes (which made use of it). Other plant cells kept both as they developed and specialized the nucleus. The original animal cells also developed the nucleus and kept the cytochromes, but abandoned the chlorophyll.

We may never know the details clearly or exactly how it all happened, but this is one rough and very speculative picture of how it came about that a billion years ago the earth had its present atmosphere, plus an ocean full of cells, both plant and animal, so that the great adventure of multicellularity, as described in Chapters 6 and 7, was ready to begin.

There seems a grim inevitability about the scheme presented in this chapter. Given a planet with the proper kind of chemistry, with a temperature that is neither too high nor too low, with an adequate air supply made up of the right gases, with an ocean, and with a sun of the right type shining down—and, most of all, given enough time!—it would seem that with grand relentlessness, first nucleic acid molecules would form, then cells, then chlorophyll (changing the atmosphere), then multicellular creatures.

Perhaps, if there is no other way of testing this scheme, we can test it by its inevitability. For instance, why does it not keep happening? Why is not life forming constantly? Why is it not forming right now in the ocean?

Ah, conditions have changed. Once nucleoproteins formed, they depleted the food supply and made less likely the independent formation of another series of nucleoproteins later on. Once the first cells appeared, then any single nucleic acid molecule that was miraculously formed was not the progenitor of a new race of life; it merely formed an article of food for some cell that blundered past. Finally, once chlorophyll started its work, oxygen filled the atmosphere, and with oxygen came the ozone layer high in the atmosphere. That meant the ultraviolet light of the sun was cut off, and that put a stop to the driving energy behind the build-up of life.

But let us look at the inevitability from another angle—

The universe is now known to be so vast, the number of stars so great, that even at worst it is hard to see how there can fail to be anything but billions upon billions of "earth-type" planets (that is, planets with environmental conditions similar to those on the earth) in the universe. Since the formation of life is inevitable according to the scheme presented here, would all of them be the habitat of some form of life? Perhaps, but there is no way we can tell yet.

We must be satisfied with those few planets we can observe, those of our own solar system, and ask if any of them would qualify as "earth-type." Unfortunately, they are a pretty bad lot from the standpoint of possible abodes of life.

Anything beyond Mars is undoubtedly too cold for life forms making use of the type of chemistry that life on the earth makes use of. Water, the necessary medium of life, exists beyond Mars only as hard-frozen and useless ice.

As for the inner planets, Mercury is too hot on one side and too cold on the other, and is airless and waterless, besides. The same, though in somewhat milder degree as far as temperature is concerned, can be said of our moon. Venus, under its eternal cloud cover, is a mystery, but recent studies seem to show its surface temperature to be above the boiling point of water, which makes it too hot for life.

That leaves Mars as the only planet that cannot be ruled out at once. Can Mars be considered "earth-type"? Perhaps, but just barely, at best.

It is smaller than the earth and has retained less of an atmosphere, one only a tenth as thick as our own. Moreover, that thin atmosphere is composed almost entirely of nitrogen, with some carbon dioxide added. Mars possesses water, but very little. There is enough to form polar ice-caps that may be a few inches thick, but it has been estimated that there is no more water on all of Mars than there is in Lake Erie. As for the temperature, the nights are of Siberian bitterness, while even the equatorial days are no warmer than a pleasant New England day. The sun, more distant from Mars than from the earth, supplies Mars with only half the ultraviolet light that the earth receives.

Surely it is asking a lot of a planet like that—thin air, practically no water, bitter cold, poor in the driving force of ultraviolet—to develop life. That is putting the inevitability of the process to a severe strain.

Well, there are dark areas on Mars that are dimly greenish in color. Plant life?

Perhaps not. It might be some form of greenish rock. However, the areas spread and contract. When it is summer in Mars' northern hemisphere, the northern icecap melts and the northern green areas expand as though growing with the increased water. Meanwhile the icecap at the South Pole grows and the green areas of the southern hemisphere contract. Half a Martian year later, the situation reverses.

Is that not how plant life would be expected to behave? Or is it a kind of rock that turns green in the presence of water and rusty red in its absence?

Is it reasonable to suppose that plant life can exist on Mars? Scientists have been trying to grow various types of bacteria, algae and lichens under conditions like those supposed to exist on Mars. Some have managed to grow. Certainly, if life forms adapted to the earth's easy conditions can manage it, then life forms adapted to harsh Mars from the start can do it easily.

And finally, in 1959, the astronomer William M. Sinton, working at the Lowell Observatory in Flagstaff, Arizona, described his studies of the light reflected from those green areas on Mars. Some of the wave lengths of sunlight were absorbed as they struck those green areas and were not reflected away in the direction of the earth's telescopes. The particular wave lengths lost were precisely those that would have been absorbed by the types of compounds found in living tissue, and not by the types found in rocks. This seems almost final proof that the green areas on Mars represent a form of life. Certainly, the only thing to be done further is to go to Mars and see (and perhaps, before too many decades have passed, some man will go to Mars and then we will find out).

But if even borderline Mars has developed life, then surely the formation of life on an earth-type planet would indeed seem to be inevitable. This would be a strong point in favor of the type of scheme presented in this chapter.

Nor is the search for life, these days, being confined to our solar system. In 1959, the astronomer, S. S. Huang, of the University of California, announced his estimation as to which of the sun's nearest neighbors might have a habitable zone of a type which could reasonably be expected to contain an earthlike planet. Two stars, Epsilon Eridani (11 light-years distant) and Tau Ceti (12 light-years away) seem the only reasonable possibilities. Both are somewhat smaller than our sun.

Otto Struve, who is heading a new radio telescope being built in West Virginia which, when completed, will be the most powerful instrument ever designed to penetrate the far reaches of space, announces that one of the tasks to which it will be put will be that of detecting radio signals that bear the signs of being originated by intelligent beings. Probably, the instrument will be pointed in the direction of Epsilon Eridani or Tau Ceti for this purpose.

We have come a long way since the question was first asked at the beginning of the book: "Where do babies come from?"

Now it is no longer sufficient to talk of storks or of

doctors' black bags, of fathers and mothers, of species, or even of cells.

Instead, to start really at the beginning, at the true wellsprings of life, we must answer something like this:

"Once upon a time, very long ago, perhaps two and a half billion years ago, under a deadly sun, in an ammoniated ocean topped by a poisonous atmosphere, in the midst of a soup of organic molecules, a nucleic acid molecule came accidentally into being that could somehow bring about the existence of another like itself—"

And from that all else would follow!

A Table of Dates

1630: Archbishop James Ussher calculates date of Creation as 4004 B. C.

1660: John Ray begins to classify plant species.

1665: Robert Hooke discovers cells in cork slices.

1668: Francesco Redi proves maggots do not arise by spontaneous generation.

1675: Anton van Leeuwenhoek discovers protozoa.

1677: Johann Ham discovers spermatozoa.

1680: Van Leeuwenhoek discovers yeast to be a microorganism.

1683: Van Leeuwenhoek discovers bacteria.

1693: Ray begins to classify animal species.

1737: Carolus Linnaeus publishes *Systema Naturae;* establishes modern science of taxonomy.

1752: René de Réaumur discovers stomach juices dissolve meat.

1767: Lazzaro Spallanzani proves microorganisms will not arise in broth that has been boiled and sealed from air.

1770: Charles Bonnet suggests periodic catastrophes have overwhelmed the earth.

1781: Felice Fontana discovers cell nuclei.

1785: James Hutton publishes *Theory of the Earth*; establishes modern science of geology.

1791: William Smith shows rock strata to contain characteristic fossils; establishes modern science of paleontology.

1791: A "sport" (short-legged sheep) is put to use for the first time.

1798: Thomas Malthus publishes *An Essay on the Principle of Population,* expounding his theories of overpopulation.

1800: Georges Cuvier popularizes "catastrophism" as a geologic theory.

1803: John Dalton proposes modern atomic theory.

1807: Jöns Berzelius divides substances into "organic" and "inorganic"; upholds "vitalism."

1809: Jean de Lamarck publishes *Zoological Philosophy*, advances theory of evolution through inheritance of acquired characteristics.

1811: Claude Berthollet discovers nitrogen in albuminous substances.

1811: Michel Chevreul breaks lipids down to fatty acids.

1812: Gottlieb Kirchhoff finds starch to be built up of glucose units.

1812: Joseph Gay-Lussac establishes chemical similarity of starch, sugar and cellulose (the carbahydrates).

1819: H. Braconnot finds cellulose to be built up of glucose units.

1820: Braconnot isolates glycine and leucine (first amino acids to be discovered) from gelatin.

1822: Johann Döbereiner discovers catalytic properties of platinum.

1824: William Prout discovers stomach juices to contain hydrochloric acid.

1827: Karl Von Baer discovers mammalian ova.

1827: Von Baer discovers notochord in mammalian embryo.

1828: Friedrich Wöhler synthesizes urea, destroys "vitalism"; establishes modern science of organic chemistry.

1829: Martin Rathke discovers mammalian embryos pass through a gilled stage.

1830: Charles Lyell publishes first volume of *Principles of Geology;* backs Hutton and destroys "catastrophism."

1831: Charles Darwin leaves on the voyage of the "Beagle."

1835: Theodor Schwann suggests that a digestive catalyst (pepsin) in stomach juices digests meat.

1836: Schwann proves microorganisms will not arise if sterilized broth is exposed to heated air.

1836: Berzelius summarizes early knowledge of catalysis; suggests use of the word "catalyst."

1838: Malthus' book inspires Darwin to work out the theory of natural selection.

1838: Schwann suggests ovum is a single cell.

1838: Gerald Mulder discovers sulfur in albuminous substances; suggests name "protein" for them.

1839: Matthias Schleiden suggests that all plants are composed of cells.

1839: Schwann suggests that all animals are composed of bells; Schleiden and Schwann thus establish "cell theory."

1840: Darwin publishes *Zoology of the Voyage of the "Beagle."*

1840: Hugo Von Mohl suggests name "protoplasm" for cell contents.

1841: Rudolf Von Kölliker suggests spermatozoon is a single cell.

1845: Karl Von Siebold suggests protozoa are single cells.

1846: Justus Von Liebig isolates tyrosine (third amino acid to be discovered) from cheese protein.

1850: Colored stains first used on cells.

1856: Claude Bernard finds glycogen to be built up of glucose units.

1857: Louis Pasteur proves fermentation of fruit juice to be brought about by living yeast cells.

1858: Friedrich Kekule begins working out structural formulas.

1858: Alfred Wallace works out theory of evolution by natural selection independently of Darwin.

1859: Darwin publishes *Origin of Species*.

1860: Pasteur finally disproves theory of spontaneous generation; all life comes from previously existing life.

1860: Rudolf Virchow summarizes cell theory; all cells come from previously existing cells.

1862: Pasteur advances germ theory of disease.

1863: Lyell publishes *The Antiquity of Man* supporting Darwinism.

1863: Thomas Huxley publishes *Man's Place in Nature* supporting Darwinism.

1866: Gregor Mendel publishes his theories of genetics; attracts no attention.

1866: Alexander Kovalevski discovers notochord in amphioxus.

1866: Ernst Haeckel points out that embryos recapitulate, during their development, the course of the evolution of the organism.

1869: Friedrich Miescher discovers nucleic acids.

1871: Darwin publishes *The Descent of Man,* suggesting evolution of the organism.

1876: Willy Kühne suggests name "enzyme" for organic evolution of man frow lower forms.

1879: Hermann Fol first observes an ovum in the process of fertilization by a single spermatozoon.

1879: Walther Flemming discovers chromatin by staining techniques.

1882: Flemming publishes *Cell-Substance, Nucleus and Cell-Division,* describing course of mitosis.

1884: Karl Von Nägeli proposes theory of orthogenesis and suggests evolution by sudden, comparatively large, jumps.

1886: Hugo de Vries comes across the first evidence out of which he works out his theory of mutations.

1888: Eduard Strassburger describes sex cells of plants as having half the number of chromosomes contained in other cells of the organism.

1889: Martinus Beijerinck uses the name "virus" to express infectious agent of tobacco-mosaic disease.

1891: Emil Fischer works out structural formulas of simple sugars.

1892: D. Ivanovski demonstrates that tobacco-mosaic virus can pass through a filter fine enough to hold back any particles visible in a microscope.

1897: Eduard Buchner shows yeast enzymes to exist and do their work even though yeast cells are killed.

1900: De Vries and two other botanists rediscover Mendel and his theories.

1902: W. S. Sutton suggests that chromosomes control the inheritance of physical characteristics.

1906: Thomas Morgan begins use of *Drosophila* in genetic experiments.

1910: James Herrick first describes sickle-cell anemia.

1911: Casimer Funk suggests the name "vitamine" for organic substances necessary to life in trace amounts.

1911: Phoebus Levene discovers ribose in one type of nucleic acid, deoxyribose in another.

1911: Peyton Rous discovers tumor virus.

1915: Felix D'Hérelle discovers bacteriophage.

1916: H. A. Allard shows that tobacco-mosaic virus can be held back by filters finer than those used by Ivanovski.

1917: First chromosome maps of *Drosophila* worked out.

1918: Fischer works out method by which amino acids join together to form proteins.

1924: Karl Freudenberg shows living tissue to contain only L-amino acids.

1925: Scopes trial in Tennessee involving the question of teaching of evolutionary theories in public schools.

1926: James Sumner crystallizes enzyme (urease) for the first time.

1927: Hermann Muller begins to expose *Drosophila* to radiation to increase incidence of mutation.

1930: John Northrop crystallizes additional enzymes.

1935: Wendell Stanley crystallizes virus (tobacco-mosaic virus) for the first time.

1939: Nucleic acids shown to be large molecules.

1941: George Beadle begins work with *Neurospora,* establishing chemical genetics.

1944: O. T. Avery discovers that genes may consist of pure nucleic acid.

1944: A. J. P. Martin describes technique of paper chromatography.

1949: Linus Pauling discovers abnormal hemoglobins and studies their inheritance.

1949: Erwin Chargaff shows various nucleotides to be present in nucleic acids in unequal proportions, but shows adenine and thymine concentrations to be equal and guanine and cytosine concentrations to be equal.

1952: S. L. Miller produces amino acids from simple compounds under primordial conditions.

1953: F. H. C. Crick and J. D. Watson advance double-strand theory of nucleic acid replication.

1954: Nucleic acid established as sole infective agent in virus.

1955: H. Fraenkel-Conrat separates virus into protein and nucleic acid, then reconstitutes virus.

1955: S. Ochoa isolates enzyme involved in RNA replication.

1956: A. Kornberg isolates enzyme involved in DNA replication.

1957: Number of human chromosomes established as forty-six per cell.

1958: R. S. Schweet produces hemoglobin via the appropriate nucleic acid in a test tube.

1958: V. G. Allfrey and A. E. Mirsky substitute synthetic polymer for nucleic acid and produce protein.

1959: William M. Sinton produces strong spectroscopic evidence in favor of the existence of plant life on Mars.

1961: M. W. Nirenberg and J. H. Matthaei begin to break the genetic code.

Index